AF835052

Resilient Safety in Advanced Air Mobility

This book explores the advanced air mobility (AAM) ecosystem, a subject that covers technological innovation, sustainable planning, and aviation safety. As urban and regional populations grow and traditional transportation networks reach their limits, AAM has emerged as a transformational solution for delivering transit through aerial mobility.

The main focus of the title is to examine how resilient safety principles can be integrated into AAM systems to create efficient, sustainable, and safe modes of local transportation. Addressing the technical, operational, and regulatory challenges of AAM, the author investigates where safety, technology, and human experience intersect, suggesting that the AAM industry is not only innovative but also safe and sustainable for all, and examines risk management strategies, showcasing the importance of embedding resilience into AAM operations to mitigate potential hazards. Technological advancements are highlighted in case studies from aircraft design innovations to vertiport infrastructure, alongside a detailed analysis of the regulatory frameworks governing the safe integration of AAM into urban settings.

The reader will gain insights and appreciate the practical guidance needed for AAM. *Resilient Safety in Advanced Air Mobility: A Framework for Sustainable Operations* is an ideal purchase for occupational health and safety practitioners, aviation professionals, systems engineers, urban planners, policymakers, and researchers.

Resilient Safety

Series Editor: Manolis Karachalios

Through a multidisciplinary lens, this series compiles and synthesizes cutting-edge research, best practices, and innovative methodologies from around the globe, encouraging the adoption of resilience-building measures across the highlighted sectors. Resilient Safety paves the way for a future where systems are not only shielded against threats but are also adaptable and robust, ensuring the well-being of societies in an ever-evolving risk landscape. Each volume within the series will embark on a detailed exploration of resilience as it pertains to its specific area of focus, from the engineering marvels that ensure the safety of aerospace, aviation, and maritime operations to the medical and public health innovations designed to withstand and respond to health crises. The series explores the intricacies of environmental protection efforts that mitigate the impact of climate change, the cybersecurity defenses that protect critical information infrastructure, and the emergency management strategies that prepare communities for disaster response and recovery. Written for a diverse readership, from industry experts and academics to policy makers and the engaged public, the books in the Resilient Safety series bridge knowledge gaps and foster a shared understanding of resilience principles.

Air Show Performers
Safety, Risk Management, and Psychological Factors
Manolis Karachalios and Daniel Kwasi Adjekum

Excellence in Air Show Performers
Training for Resilient Safety
Manolis Karachalios

Resilient Safety in Advanced Air Mobility
A Framework for Sustainable Operations
Manolis Karachalios

For more information about this series, please visit: https://www.routledge.com/Resilient-Safety/book-series/EIERS

Resilient Safety in Advanced Air Mobility

A Framework for Sustainable Operations

Manolis Karachalios

CRC Press
Taylor & Francis Group
Boca Raton London New York

CRC Press is an imprint of the
Taylor & Francis Group, an informa business

Designed cover image: Drone group control system illustration by Tuzla/Shutterstock

First edition published 2026
by CRC Press
2385 NW Executive Center Drive, Suite 320, Boca Raton FL 33431

and by CRC Press
4 Park Square, Milton Park, Abingdon, Oxon, OX14 4RN

CRC Press is an imprint of Taylor & Francis Group, LLC

© 2026 Manolis Karachalios

Reasonable efforts have been made to publish reliable data and information, but the author and publisher cannot assume responsibility for the validity of all materials or the consequences of their use. The authors and publishers have attempted to trace the copyright holders of all material reproduced in this publication and apologize to copyright holders if permission to publish in this form has not been obtained. If any copyright material has not been acknowledged please write and let us know so we may rectify in any future reprint.

With the exception of Chapter 1, no part of this book may be reprinted or reproduced or utilised in any form or by any electronic, mechanical, or other means, now known or hereafter invented, including photocopying and recording, or in any information storage or retrieval system, without permission in writing from the publishers

Chapter 1 of this book is freely available as a downloadable Open Access PDF at http://www.taylorfrancis.com under a Creative Commons [Attribution-Non Commercial-No Derivatives (CC BY-NC-ND)] 4.0 license.

For permission to photocopy or use material electronically from this work, access www.copyright.com or contact the Copyright Clearance Center, Inc. (CCC), 222 Rosewood Drive, Danvers, MA 01923, 978-750-8400. For works that are not available on CCC please contact mpkbookspermissions@tandf.co.uk

For Product Safety Concerns and Information please contact our EU representative GPSR@taylorandfrancis.com. Taylor & Francis Verlag GmbH, Kaufingerstraße 24, 80331 München, Germany.

Trademark notice: Product or corporate names may be trademarks or registered trademarks and are used only for identification and explanation without intent to infringe.

ISBN: 978-1-032-89774-5 (hbk)
ISBN: 978-1-032-90141-1 (pbk)
ISBN: 978-1-003-54630-6 (ebk)

DOI: 10.1201/9781003546306

Typeset in Times
by KnowledgeWorks Global Ltd.

To our adored daughters, Maria and Dimitra,
who oxygenate our life.

Contents

PART IV The AAM Outlook

About the Author

Manolis Karachalios (PhD, MBA, ATP, CFI, FRAeS) is a visionary leader in aviation with over 25 years of experience in global operations and strategic management. Renowned for leading multinational teams and pioneering safety innovations, he was the Hellenic Air Force's F-16 Demo Team "Zeus" Display Pilot, currently a Board Member of the European Airshow Council, and Flying Display Director for the Athens Flying Week International Air Show, while served as Senior Lecturer in the Business Management Department at Emirates Aviation University. Holding an MBA from Coventry University and a PhD in Aerospace Sciences from the University of North Dakota, Manolis authored *Excellence in Air Show Performers: Training for Resilient Safety* and coauthored *Air Show Performers: Safety, Risk Management, and Psychological Factors*. Nowadays, he has shifted his focus in promoting a framework for the sustainable evolution of the advanced air mobility ecosystem, which he considers the incubator of Aviation 5.0.

Preface

Like many of you, I grew up enchanted by the captivating visions of the future portrayed in popular culture. As a child, I vividly remember sitting wide-eyed in front of the television, watching the animated world of The Jetsons[1]. I admired at how effortlessly George Jetson navigated bustling skylines in his flying vehicle. It was an inspiring glimpse into a future where people could rise above the constraints of traditional mobility, traveling swiftly and seamlessly through the air. At the time, this futuristic reality seemed fantastical—an impossible yet tantalizing dream born purely of human imagination.

Decades later, that childlike fascination with aerial mobility still fuels my curiosity, and today, the dream once depicted by The Jetsons is rapidly transitioning into reality. Advanced air mobility (AAM)—encompassing electric vertical takeoff and landing (eVTOL) aircraft, urban air taxis, autonomous aerial vehicles, and drone-based logistics—is no longer just the realm of science fiction. It is actively emerging as a transformative technology that holds profound implications for society, infrastructure, and our everyday lives.

Yet, as promising as this airborne revolution is, it also comes with formidable challenges. My journey as an aviation safety researcher and practitioner has afforded me firsthand insights into the incredible possibilities and equally daunting complexities associated with bringing advanced air vehicles into widespread use. One fundamental truth has become increasingly clear to me over the years: Innovation, no matter how inspiring, cannot thrive without an uncompromising commitment to resilient safety and sustainable operations.

This realization deeply motivated me, prompting the conception of this book: *Resilient Safety in Advanced Air Mobility: A Framework for Sustainable Operations*. My intention is to provide you—whether you're an aviation professional, researcher, regulator, policymaker, student, or simply someone deeply intrigued by the promise of future transportation—with a clear and practical guide to understanding the intricate interplay between innovative technology, safety culture, and operational sustainability. I hope that together, we will explore how resilient safety principles and thoughtful operational planning can enable AAM to fulfill its true potential while proactively mitigating the unique risks inherent in pioneering aviation frontiers.

Throughout the following pages, we will systematically examine the multifaceted ecosystem underpinning AAM by addressing four key questions (which coincide with the four parts that this publication is divided into): *why* we need AAMs, *how* we can incorporate resilient safety principles into them, *what* we actually achieve by doing so, and *where* we go next. We will explore technical, regulatory, human and automation-centric aspects, considering everything from innovative aircraft design and certification pathways to infrastructure integration, public perception, and community engagement. We will discuss proven methods to manage emerging risks, adapt dynamically to unforeseen disruptions, and foster the culture of trust and reliability needed to ensure AAM's long-term viability and public acceptance.

Ultimately, my hope is that this book could contribute to the resilient, sustainable, and exponential growth of aviation via AAM.

NOTE

1. The Jetsons is an American Hanna-Barbera animated sitcom (1962–1963, revived 1985–1987) that projects a suburban nuclear family into a gadget-laden Space Age future to satirize mid-century techno-optimism (Graydon, 2011).

REFERENCE

Graydon, D. (2011). *The Jetsons: The official guide to the cartoon classic*. Running Press.

Acknowledgments

This book couldn't have become a reality without a profound impact, indirectly, by studying the work of professors Eric Hollnagel, James Reason, and Sidney Dekker over the last several years. Their contribution to safety science is phenomenal, multifaceted, and truly inspiring to me for the principles they suggest of the resilient engineering, the *Swiss cheese model*, and the *new safety view*. Their work can be easily adapted to most of the modern industries, and advanced air mobility can't be excluded. Once again, I'm grateful to Dr. Daniel Kwasi Adjekum, who introduced me to all the above forward-thinking ideas, as well as to the concept of resilient safety, which is the focal point of the current publication.

List of Acronyms

AAM	Advanced Air Mobility
ALARP	As low as reasonably practicable
AAT	Autonomous air taxi
AAV	Autonomous Aerial Vehicle
AC	Advisory Circular—Regulatory guidance (FAA/EASA)
ADS-BADS-B	Automatic Dependent Surveillance–Broadcast
AFACS	Autonomy Factors Analysis and Classification System
AFR	Autonomous Flight Rules
AGL	Above Ground Level—Altitude reference
AI/ML	Artificial Intelligence/Machine Learning
ARB	Air risk buffer
ASTM	ASTM International (consensus standards; includes UAS committees)
ATC	Air Traffic Control
ATM	Air Traffic Management
BVLOS	Beyond Visual Line of Sight
C2	Command and Control—UAS command link and control function
CAAC	Civil Aviation Administration of China
CAST	China Academy of Civil Aviation Science and Technology
CCC	Command and Control Center
CCS	Combined Charging System (Charging Standard)
CRM	Crew Resource Management—Teamwork/communication for safety
CV	Contingency volume
DAA	Detect and Avoid—Functions/sensors to remain well clear and avoid collisions
DAL	Design Assurance Level (A–E)—Integrity level for airborne systems
DME	Distance Measuring Equipment
EASA	European Union Aviation Safety Agency (EU regulator)
EMS	Emergency Medical Services
ECG	Electrocardiogram
EEG	Electroencephalogram
eVTOL	Electric Vertical Takeoff and Landing
eSTOL	Electric Short Takeoff and Landing
FAA	Federal Aviation Administration (U.S. regulator)
FG	Flight geography
FIMS	Flight Information Management System
FMEA/FMECA	Failure Modes and Effects (and Criticality) Analysis
FRAM	Functional Resonance Analysis Method
FRAT	Flight risk assessment tool

FTS	Final termination system
GNSS	Global Navigation Satellite System (e.g., GPS, Galileo, GLONASS, BeiDou)
GRB	Ground risk buffer
HAT	Human-autonomy teaming
HF	Human Factors—Ergonomics, cognition, workload, etc.
HFACS	Human Factors Analysis and Classification System
HIL	Hardware in the Loop—Real hardware + simulated environment testing
HIRA	Hazard Identification and Risk Assessment
HITL	Human-In-The-Loop
HMI	Human–Machine Interface—Interaction surfaces between humans and systems
HRO	High Reliability Organization
ICAO	International Civil Aviation Organization (UN specialized agency)
IFR/VFR	Instrument/Visual Flight Rules
LAANC	Low Altitude Authorization and Notification Capability—Near real time UAS airspace authorizations
LORAN	LOng RAnge Navigation (predecessor to the modern GPS systems).
LiDAR	Light Detection and Ranging
NAS	National Airspace System—U.S. airspace, facilities, procedures, and users
RAG	Resilience Analysis Grid
RID	Remote ID—Broadcast/Network identification of UAS in flight
RSO	Remote supervisor operator
RTH	Return To Home
RTL	Return To Land
SAF	Sustainable Aviation Fuel
SAIL	Specific Assurance and Integrity Level—Assurance level within SORA
SMS	Safety Management System—Organization-level safety framework
SORA	Specific Operations Risk Assessment (JARUS methodology)
STPA	System Theoretic Process Analysis—Control theoretic hazard analysis
TC	Type Certificate—Approval of aircraft/engine/propeller design
TCAS	Traffic Collision Avoidance System
UAM	Urban Air Mobility—AAM focused on dense urban environments
UAS	Unmanned Aircraft System—Aircraft + control station + communications link
UAV	Unmanned Aerial Vehicle—The aircraft component of a UAS
UML	UAM Maturity Levels
UOMS	UAS Operation Management System

USS	UAS Service Supplier—Provider interfacing operators to UTM services
UTM	UAS Traffic Management—Services for low-altitude UAS operations
VLOS	Visual Line of Sight
V&V	Verification and Validation
VOR	VHF Omnidirectional Range
WAI	Work as Imagined
WAD	Work as Done

Part I

The AAM Ecosystem

Looking at the historical review of Chapter 1, one will sense the factors that have shaped the vision of advanced air mobility (AAM) and propelled it from a futuristic idea to a developing reality. Yet, AAM's foundations are driven by a clear *why*: a global need for cleaner, faster, and more equitable transportation solutions, balanced by the understanding that successful adoption rests on prudent integration and foresight in design.

Today's transportation systems are at a crossroads. Cities worldwide grapple with gridlocked roads and constrained infrastructure, while many rural and suburban regions remain disconnected from major transport networks. Traditional solutions like expanding highways or rail lines are often impractical or unsustainable, prompting urban planners to champion concepts such as the "*15-minute city*" to minimize long commutes and emissions (Khavarian-Garmsir et al., 2023; Pozoukidou & Chatziyiannaki, 2021). Yet these initiatives alone cannot bridge all mobility gaps or address the equity issues faced by underserved communities. Meanwhile, the climate imperative to decarbonize mobility grows stronger as nations set ambitious net-zero targets. In this context, AAM has emerged as a promising answer to *why* our transport paradigm must evolve, leveraging the vertical dimension to overcome ground-level limits and connect people in new, efficient ways.

AAM refers to next-generation aerial transport systems—from electric vertical takeoff and landing (eVTOL) air taxis to autonomous cargo drones—designed to move people and goods in spaces previously beyond the reach of regular aviation (e.g., across metro areas or to remote villages) (Drévillon, 2024). By tapping underutilized airspace, AAM can bypass traffic bottlenecks and leapfrog geographic barriers, dramatically improving point-to-point travel times while relieving pressure on existing roads (Walker et al., 2024). Crucially, these aircraft are typically electric-powered,

DOI: 10.1201/9781003546306-1

producing zero operational emissions; this supports urban decarbonization efforts and aligns with aviation's broader push toward net-zero carbon by mid-century. Equally important is AAM's potential to enhance connectivity and equity: it can link isolated communities to essential services and economic opportunities by delivering medical supplies, enabling emergency response, or simply providing faster access to cities for those in remote areas. Industry roadmaps and academic studies alike reflect a cautious optimism—AAM is widely seen as a transformative mobility innovation that could boost sustainability and inclusivity, *provided* it is integrated thoughtfully. Achieving its full potential will require careful long-term planning, public engagement, and policy support to ensure these new air services complement existing transit systems and serve society as a whole.

1 AAM—Evolution

INTRODUCTION

Imagine a future commute where you hail an air taxi that lifts off vertically from a nearby vertiport and carries you above the traffic to your destination. This vision, once part only of science fiction, is rapidly becoming a serious pursuit in the aviation world. Advanced air mobility (AAM) refers to the emerging aviation ecosystem that could make such everyday flight a reality. It encompasses new types of aircraft (like electric "flying cars" and delivery drones) and new infrastructure, aiming to *bring the movement of people and goods off the ground, on demand, and into the sky* (NASA, n.d.). Aviation agencies, technology companies, and researchers are investing heavily in AAM, heralding it as the next revolution in air transport. Some forecasts predict a massive economic impact—for example, a 2021 Morgan Stanley analysis estimated the AAM market could reach around $1 trillion by 2040—though other analysts offer more modest projections and caution that these numbers depend on many unknowns (Brinkmann, 2022a; Guo et al., 2024). Hype aside, there is broad agreement that AAM has the potential to transform how we travel and transport goods, much like the advent of jet airliners or GPS navigation did in earlier eras.

WHAT EXACTLY IS ADVANCED AIR MOBILITY?

Nowadays, there is so much talk about drones, air mobility, and air taxis that, before we continue with this book, we need to answer the question: *What exactly is AAM?* At its core, AAM is about a new air transportation system for local and regional travel, enabled by cutting-edge aircraft designs and automation. NASA formally defines AAM as "*safe, sustainable, affordable, and accessible aviation for transformational local and intraregional missions*" (Garrow et al., 2022, p. 2). In simpler terms, AAM envisions using small, highly advanced aircraft to move people or cargo in ways that have not been routine before—for example, zipping above congested highways or connecting a city with outlying towns quickly. The word "*transformational*" in NASA's definition is key: the intent is not just incremental improvement, but a substantive change in the role aviation plays in our daily lives. This could mean making short-hop flights as ordinary to the public as taking a bus or an Uber is today.

AAM is often described as an *ecosystem* because it is more than just the vehicles. It includes new aircraft types, support infrastructure, airspace integration, and operational models, all working harmoniously together. The aircraft central to AAM are typically lightweight, highly automated, and capable of vertical takeoff and landing (VTOL). Many are envisioned to be electrically powered (hence the common term *eVTOL* for electric VTOL aircraft), using battery-driven motors instead

DOI: 10.1201/9781003546306-2

This chapter has been made available under a CC-BY-NC-ND license.

of conventional fuel engines. These aircraft might carry a handful of passengers or small cargo payloads, and some designs look quite futuristic—like multicopter drones scaled up to carry people, or small planes with tilting propellers that allow hovering. A variety of propulsion technologies fall under the AAM umbrella, including fully electric systems and hybrid-electric designs, and potentially even hydrogen fuel cell concepts for longer-range use. What they share is the use of *advanced technologies* (Federal Aviation Administration, 2023) such as electrification, operational autonomy, and digital automation, which distinguish them from traditional airplanes and helicopters.

Crucially, AAM is broader than the popular term *urban air mobility (UAM)* which one may have heard. UAM generally refers to air transportation services within urban areas—think air taxis shuttling across a city or drones delivering packages downtown. AAM *includes* UAM but extends beyond cities. It covers *local* missions (intra-city) as well as *intraregional* missions that might connect suburban or rural areas to a city or link two smaller cities within a region. For instance, a 20-mile hop from a downtown to a suburb would be a UAM operation, whereas a 200-mile flight connecting two cities in a state falls into the AAM vision as well (sometimes termed regional air mobility). In fact, the U.S. FAA uses *AAM* as an umbrella term encompassing both urban and regional use cases, *moving people and property by air between two points using aircraft with advanced technologies, including electric or eVTOL, in both controlled and uncontrolled airspace* (Federal Aviation Administration, 2023). The inclusion of uncontrolled airspace in that definition demonstrates that AAM operations could take place at low altitudes outside of conventional airport routes, which is a departure from most of today's commercial aviation that sticks to controlled airways and airports.

To make AAM possible, a whole support system must be developed. One aspect is physical infrastructure: networks of vertiports (small airports or helipads for VTOL craft) where passengers can board, and where aircraft can recharge or swap batteries. Cities might see rooftop or parking-lot vertiports, while regional operations might use existing small airports or new landing sites in communities. Another aspect is digital infrastructure, often referred to as UTM (uncrewed traffic management) or, more generally, air traffic management for AAM. Because the vision involves potentially large numbers of new vehicles (some without onboard pilots) sharing low-altitude airspace, traditional air traffic control methods won't suffice. Highly automated systems will likely be needed to deconflict routes, manage flight corridors, and integrate AAM traffic with existing airplanes and helicopters (NASA, n.d.; Tang et al., 2021; Xie et al., 2021). Think of it as building "*walking paths in the sky*" with designated lanes or corridors for these vehicles, and traffic management largely handled by software and networked communication between aircraft and ground systems.

Looking at the big picture, AAM is a multifaceted concept aiming to make air travel a part of everyday local transport. It brings together novel aircraft (like eVTOL air taxis and delivery drones), new infrastructure (vertiports, charging stations), and new operational frameworks (automated traffic management, app-based on-demand flight services). The ultimate goal is an air transportation network that is accessible (available to more people, not just high-end travelers), affordable (competitive in cost with other modes), safe (meeting the stringent safety levels of aviation today or better), and sustainable (reducing noise and emissions)—all in service of transforming mobility. It's an ambitious goal, and to appreciate how we got here, it helps to look back at the evolution of this idea and the technological progress that underpins it.

EVOLUTION OF AAM: A HISTORICAL OVERVIEW (1910s–2025)

AAM has roots reaching back over a century. From the earliest vertical flight experiments to today's electric air taxis, the development of AAM unfolds as a global narrative of technological trial and error, visionary ambition, and incremental breakthroughs. What follows is a chronological tour from the first *VTOL* concepts through the state of *AAM* in 2025, highlighting key milestones and case studies around the world (see Figure 1.1).

Early VTOL Experimentation (1910s–1930s)

Long before anyone imagined AAM, engineers were wrestling with vertical flight. Early attempts were crude but instructive. In France, Paul Cornu achieved a brief, unstable hover in 1907 with a twin-rotor machine (see Figure 1.2)—historically notable, even if control and power-to-weight were poor (The Editors of Encyclopaedia Britannica, n.d.-a). In the United States, Emile and Henry Berliner flew short liftoffs with coaxial-rotor prototypes (1908–1909); the hops were low, often steadied, but proved the lift concept (Smithsonian National Air and Space Museum, n.d.-a). Progress accelerated with Étienne Oehmichen, whose multirotor experiments culminated in the first FAI-recognized helicopter distance record (360 m) in 1924—critical proof that a helicopter could fly a controllable, point-to-point trajectory (Fédération Aéronautique Internationale, n.d.). Meanwhile, the U.S. Army funded George de Bothezat's giant quadrotor (1922–1923). It flew over a hundred test sorties—but rarely above a few meters, and was deemed too complex to be practical (Smithsonian National Air and Space Museum, n.d.-a). By the mid-1930s, designs crossed a threshold: Germany's Focke-Wulf Fw 61 (first flight 1936) is often credited as the first truly practical helicopter, demonstrating stable hover and forward flight; in the United States, Sikorsky's VS-300 (1939) set the lineage for the first mass-produced helicopters (The Editors of Encyclopaedia Britannica, n.d.-b). These

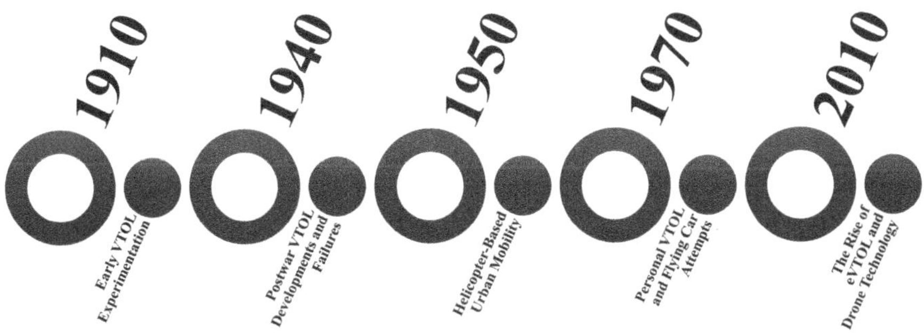

FIGURE 1.1 AAM's evolution timeline.

FIGURE 1.2 French inventor Paul Cornu with his helicopter, 1907. (*Source:* Wikimedia.)

milestones established the fundamentals—sustained vertical lift, controllability, and transition to useful flight—on which the postwar era would build.

Postwar VTOL Developments and Failures (1940s–1960s)

With helicopters maturing after WWII, militaries chased faster VTOL concepts during the 1950s–1960s—tail-sitters, tilt-wings, lift-jets, and ducted fans. Many flew, most failed. The U.S. Navy's tail-sitter fighters—Convair XFY-1 "Pogo" and Lockheed XFV-1 (1954–1955)—were airworthy but impractical to land vertically on pitching decks (Polmar, 2013a, 2013b). The Ryan X-13 Vertijet and Bell X-14 explored vectored-thrust hover; the Bell XV-3 and later NASA/Bell XV-15, see Figure 1.3, (first flight in 1977) refined the tilt-rotor pathway that ultimately led to the V-22 (NASA, 2005). Canada's CL-84 Dynavert (tilt-wing) convincingly demonstrated concept feasibility, yet never entered service after accidents and program headwinds. Among the most studied failures, the Avro Canada VZ-9 "Avrocar" (late 1950s–1961) suffered severe "hubcapping" instabilities and poor performance above ground-effect, highlighting the aero-control pitfalls of unconventional VTOL (National Museum of the U.S. Air Force, n.d.).

Europe made seminal contributions. The United Kingdom's Rolls-Royce Thrust Measuring Rig ("Flying Bedstead," first free flight in 1954) was the world's first pure jet-lift rig and informed VTOL control with reaction jets (Science Museum Group (London), n.d.). Most importantly, Hawker's P.1127 (first flight in 1960) proved vectored-thrust viability, evolving into the Harrier, the world's first operational VTOL jet fighter (service from 1969) (The Editors of Encyclopaedia Britannica, n.d.-c). France's annular-wing tail-sitter SNECMA C.450 Coléoptère crashed in 1959 during

FIGURE 1.3 NASA-Army-Bell XV-15 tiltrotor research aircraft. (*Source:* Wikimedia.)

transition trials, and the concept did not proceed. The Soviet Navy fielded the Yak-38 in the 1970s; it reached service, but with significant operational compromises. By the late 1960s, most experimental VTOL lines (outside the Harrier and helicopter evolution) had been curtailed—leaving a "museum of prototypes," but also a deep well of data on control, transition, propulsion, and pilot workload that AAM designers still mine.

Helicopter-Based Urban Mobility (1950s–1980s)

If many futuristic VTOLs struggled, helicopters did not. From the 1950s through the 1970s, several cities trialed urban helicopter airlines to beat traffic—an early precursor to AAM services. Los Angeles Airways launched scheduled passenger services in the 1950s; New York Airways shuttled between Manhattan heliports (including the Pan Am Building rooftop) and regional airports; in the San Francisco Bay Area, SFO Helicopter Airlines linked downtowns to SFO in 8–13 minutes, with fares of roughly $4–$15 (and interline subsidies in some cases). These services proved the time-savings and demand, but faced structural headwinds: high operating costs, fuel burn, maintenance intensity, and noise. Accidents—most notably the 1977 Pan Am rooftop tragedy involving a New York Airways S-61L—eroded public acceptance and economics; most operators ceased by the late 1970s (Belinn, 1956; NTSB, 1977; SFGate, 2025). Even so, these programs left infrastructure (e.g., city heliports) and a trove of operational lessons—vertiport siting, community noise management, and safety protocols—that inform today's eVTOL integration planning.

Personal VTOL and Flying Car Attempts (1970s–2000s)

The post-Apollo imagination fueled repeated attempts at personal VTOL: jet belts, flying platforms, ducted fan "saucers," and roadable aircraft. Some worked in narrow envelopes. The Bell Rocket Belt (1960s) delivered spectacular seconds-long hops but was operationally impractical due to vanishing endurance (Smithsonian National Air and Space Museum, n.d.-b). The Williams WASP/X-Jet (1970s–1980s), a single-operator ducted-fan "flying pulpit," offered much better endurance and controllability but still lacked compelling military or civil use cases once small helicopters and (later) unmanned systems advanced; the Army did not adopt it. Entrepreneurial projects—most famously Paul Moller's decades-long Skycar quest—demonstrated persistent vision and considerable ingenuity but never crossed into certified, practical products. In hindsight, these efforts were technology-pathfinders: they highlighted the control-system, energy-density, reliability, and certification gaps that later eVTOL architectures (and battery progress) would have to close.

The Rise of Electric VTOL and Drone Technology (2010s)

The 2010s supplied the missing ingredients for practical VTOL in short-range missions: lithium-ion batteries with improved specific energy and power; high-efficiency electric motors; inexpensive MEMS sensors and fly-by-wire; and—critically—drone autonomy and control software matured at scale. Rigorous, peer-reviewed analyses showed that eVTOL missions of tens of miles are energetically plausible with then-current batteries, while also quantifying binding constraints (reserve energy, turnaround, cycle life) (Fredericks et al., 2018; Sripad & Viswanathan, 2021). In parallel, policy and ecosystem work accelerated. Uber Elevate's 2016 white paper articulated a network-level vision (skyports, target performance, ops concepts) and galvanized industry/regulator focus (Holden & Goel, 2016). NASA's Urban Air Mobility Grand Challenge established a structured, scenario-based approach to safety and scalability and identified research gaps (airspace integration, procedures, certification enablers) (NASA, 2020a, 2020b). EASA (2019) issued SC-VTOL-01, a first-of-kind certification framework for VTOL-capable aircraft, and the FAA expanded UAM/AAM ConOps and vertiport guidance—indicating regulatory traction even as standards evolve. On the vehicle side, startups and incumbents fielded full-scale demonstrators: Volocopter (multirotor), Joby (tilt-prop), Archer, Lilium (ducted fan "jet"), Beta (eCTOL/eVTOL), Eve (Embraer), and CityAirbus. High-visibility demos, e.g., Dubai's 2017 public autonomous air taxi flight with Volocopter, moved the conversation from renderings to flight (Volocopter/Dubai RTA, 2017). By 2023–2024, EHang secured the world's first type certificate for a pilotless, passenger-carrying eVTOL (EH216-S) from China's Civil Aviation Administration of China (CAAC), followed by production approvals—material signals that certification is possible, even if regional paths differ (CAAC/EHang, 2023).

Technically, the eVTOL value proposition—lower noise, simpler drivetrains, digital stability, and zero local emissions—addresses the core weaknesses of helicopter-era urban mobility. The remaining constraints are well characterized: battery-specific energy and cycle life, infrastructure/airspace integration, and certification/operations at scale (NASA, 2019; Sripad & Viswanathan, 2021). The 2010s, in short, turned VTOL's century of "almosts" into a credible AAM roadmap for the 2020s.

Milestones in the 2020s: Global Case Studies

In the first half of the 2020s, AAM initiatives transitioned from prototype to early implementation. Key milestones around the world (see Figure 1.4) signaled the dawn of the air taxi era. By 2025, the United States, China, and the United Arab Emirates stood out as hotspots of AAM activity, thanks to supportive policies and flagship projects.

United States

California remains the gravitational center for eVTOL development (e.g., Joby, Archer, Wisk). In June 2023, Joby Aviation received an FAA Special Airworthiness Certificate for its first production-line prototype, permitting flight testing (a limited, non-commercial authorization) (Joby Aviation, 2023a; Molyneaux, 2023). In September 2023, Joby delivered its first aircraft to the U.S. Air Force at Edwards AFB under AFWERX Agility Prime, initiating on-base operations—an important *government-as-first-customer* pathway prior to civil service (Joby Aviation, 2023b; U.S. Air Force, 2023). Through 2024–2025, Joby reported continued progress into Type Inspection Authorization (TIA) testing, the final FAA phase before type certification (CompositesWorld, 2023).

In parallel, Archer and United Airlines announced a planned EWR ↔ Downtown Manhattan route (subject to certification and infrastructure readiness). Announcements in late 2022 set an initial target around the mid-2020s; subsequent coverage continues to position the route as an early U.S. use case, contingent on FAA approvals and vertiport arrangements (Archer Aviation, 2022; Brinkmann, 2022b; Hawkins, 2025).

On the ecosystem side, NASA's Advanced Air Mobility National Campaign has conducted surrogate eVTOL flight tests (e.g., Bell 407GXi) and acoustic campaigns with full-scale AAM aircraft to generate data for airspace integration, automation concepts, and community noise (Holder & Mitchell, 2023; Whiting, 2022). These activities are prerequisites for scaling AAM operations even as type certification proceeds.

China

China has advanced rapidly on regulatory approvals and public demonstrations, particularly via EHang. In October 2023, the CAAC issued a type certificate for EHang's EH216-S (pilotless, passenger-carrying eVTOL)—the first of its kind globally (Butterworth-Hayes, 2023; EHang Holdings Limited, 2023a). In 2024, EHang also obtained a Production Certificate from CAAC, followed by standard airworthiness approvals for delivered aircraft—formal steps enabling scaled manufacturing and supervised deployments (EHang Holdings Limited, 2024; Goldstein, 2024). EHang then conducted commercial demonstration flights carrying officials and invited passengers in Guangzhou and Hefei (Daleo, 2023; EHang Holdings Limited, 2023b).

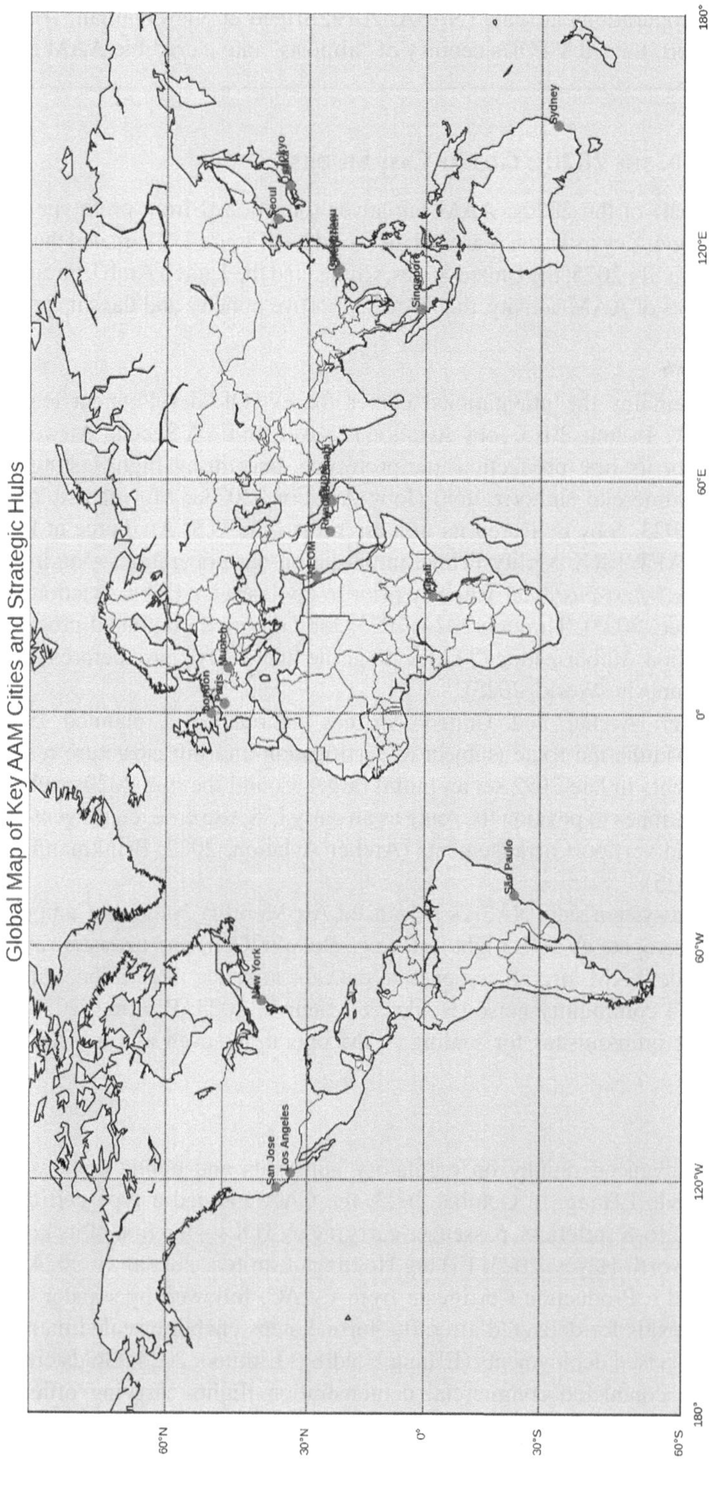

FIGURE 1.4 Global AAM distribution. (*Source:* Author.)

Beyond passenger demos, China's *low-altitude economy* push is visible in logistics. Meituan has operated thousands of drone delivery sorties in Shenzhen (see Figure 1.5) and elsewhere; by April 2025, it reported a nationwide low-altitude logistics operating license, illustrating how cargo missions can seed AAM operations frameworks (*China Daily*, 2025).

Another Chinese developer, AutoFlight, demonstrated improving battery-range performance. In February 2023, its Generation-4 Prosperity eVTOL (lift-plus-cruise) flew 250.3 km on a single charge under remote piloting, a publicly verified long-distance record for the class at the time—useful evidence for near-term inter-city missions (AutoFlight, 2023; U.S. Air Force/AFWERX (via *Avionics International*), 2022).

FIGURE 1.5 Meituan UAVs in Shenzhen. (*Source:* Author.)

United Arab Emirates

The UAE has positioned AAM within a broader smart-mobility strategy. In February 2024, Dubai's Roads and Transport Authority (RTA) signed a definitive agreement granting Joby exclusive rights to operate air-taxi services in Dubai for six years, alongside Skyports for vertiports—a citywide commitment to AAM infrastructure and operations (Joby Aviation, 2024; Roads and Transport Authority [RTA] Dubai, 2024). In June and July 2025, Joby conducted piloted, full-transition eVTOL flights in Dubai, and construction began on a DXB vertiport targeting Q1 2026 completion—milestones indicative of pre-commercial "market readiness" work (Joby Aviation, 2025; Ros, 2025). Moreover, drone delivery by Keeto Drones commenced in Dubai in 2025, showcasing the emirate's vision to establish as a fore fronter in the evolution of new technologies (Varghese, 2025).

Abu Dhabi has moved in parallel, partnering with Archer. In July 2025, Archer completed an initial flight of its Midnight eVTOL at Al Bateen Executive Airport, with UAE stakeholders present—another step toward localized environmental and operations validation (Business Wire, 2025; Stonor, 2025). In addition, Abu Dhabi has showcased autonomous EH216-S demo flights, signaling regulatory interest in both piloted and pilotless architectures (TOI World Desk, 2025).

The next emirate within the UAE to embrace AAMs was the most northern one: Ras Al Khaimah (RAK). In September 2025, a set of test flights of XPeng AeroHT's took place at Jazeerah Aviation Club airport with the presence of the RAK's ruler, showcasing their intent to offer inter-emirate connections from RAK to Dubai, as well as within the emirate for tourism experiences, but also local transport and even emergency response in a faster, convenient, and eco-friendly means (wow-RAK, 2025).

Other Regions

Paris sought to showcase eVTOLs during the 2024 Olympics, but passenger operations were not authorized in time; instead, crewed test flights proceeded at validation sites as part of the operational trials (Ridden, 2024; Vidalon, 2024). Airbus advanced CityAirbus NextGen from unveiling to initial liftoff/hover by late 2024 as part of an ongoing test program (Airbus, 2024; Aerospace Global News, 2024). On the regulatory side, EASA issued updated Means of Compliance for SC-VTOL in July 2025, reflecting a maturing certification framework (EASA, 2025a; EASA, 2025b).

In Japan, SkyDrive began demonstration flights with its SD-05 at Expo 2025 Osaka, consistent with Japan's staged AAM deployment goals (Bielby, 2025; SkyDrive, Inc. 2025).

Volocopter completed South Korea's first crewed public eVTOL test flight in Seoul in November 2021, a government-backed step in the country's *K-UAM* roadmap (Volocopter, 2021), while the company also executed a manned Marina Bay flight in October 2019 under CAAS supervision (Volocopter, 2019).

REFERENCES

Archer Aviation Inc. (2022, November 10). *Archer and United Airlines announce first commercial electric air taxi route in the US: Downtown Manhattan to Newark Liberty International Airport* [Press release].

AutoFlight. (2023, March 2). AutoFlight claims record for longest distance eVTOL flight [Press release]. *Vertical Magazine.*

Belinn, C. M. (1956). Los Angeles Airways—The world's first scheduled helicopter airline. *Journal of the American Helicopter Society, 1*(1), 111–117. https://doi.org/10.4050/JAHS.1.111

Bielby, K. (2025, April 9). *SkyDrive demonstrates SKYDRIVE SD-05 at Osaka-Kansai Expo. Urban Air Mobility News.*

Brinkmann, P. (2022a, August 25). *Making sense of advanced air mobility market projections. Aerospace America* (AIAA).

Brinkmann, P. (2022b, November 10). *Archer, United Airlines choose Newark-to-Manhattan route as the first for Archer's air taxis. Aerospace America* (AIAA).

Business Wire. (2025, July 2). Archer begins test flights in Abu Dhabi. *Business Wire.*

Butterworth-Hayes, P. (2023, October 13). *EHang announces CAAC type certification for its EH216-S autonomous air vehicle. Unmanned Airspace.*

CAAC/EHang. (2023, October 13; 2023–2024 updates). EH216-S type certification and production certification announcements. *EHang & CompositesWorld.* https://www.ehang.com/news/1058.html; https://www.compositesworld.com/news/ehang-obtains-eh216-s-evtol-type-certification-from-caac

China Daily. (2025, April 22). Meituan secures first nationwide low-altitude logistics operating license. *China Daily.*

Clean Air Task Force (Walker, T. K. III, Tatsutani, M., & Lewis, J.). (2024, April 30). *Decarbonizing aviation: Enabling technologies for a net-zero future.* https://cdn.catf.us/wp-content/uploads/2024/04/03083920/decarbonizing-aviation-technologies-net-zero-future.pdf

CompositesWorld. (2023, December 21). EH216-S eVTOL aircraft gains CAAC standard airworthiness certification. *CompositesWorld.*

Daleo, J. (2023). *EHang makes world's first commercial electric air taxi flight.* Flying Magazine.

Drévillon, H. (2024, July 2). Integrating advanced air mobility in urban environments. *Egis Group.* https://www.egis-group.com/all-insights/integrating-advanced-air-mobility-in-urban-environments

EHang Holdings Limited. (2023a, October 13). *EHang successfully obtains type certificate for EH216-S passenger-carrying UAV system issued by Civil Aviation Administration of China* [Press release]. https://www.ehang.com/news/1058.html

EHang Holdings Limited. (2023b, December 28). *EHang's certified EH216-S pilotless passenger-carrying aerial vehicles debut commercial flight demonstrations in Guangzhou and Hefei.* EHang News Center.

EHang Holdings Limited. (2024, April 7). *EHang secures production certificate from CAAC, clearing path for mass production of EH216-S pilotless eVTOL aircraft.* EHang News Center. https://www.ehang.com/news/1058.html

European Union Aviation Safety Agency. (2019, July 2). *Special Condition VTOL: Special Condition for small-category VTOL aircraft* (Doc. No. SC-VTOL-01, Issue 1).

Federal Aviation Administration. (2023). Advanced Air Mobility (AAM) implementation plan: Near-term (Innovate28) focus with an eye on the future of AAM (Version 1.0). *FAA.* https://www.faa.gov/sites/faa.gov/files/AAM-I28-Implementation-Plan.pdf

Fédération Aéronautique Internationale. (n.d.). Historical helicopter records (Étienne Oehmichen, 1924—distance 360 m, etc.). *FAI.*

Fredericks, W. L., Sripad, S., Bower, G., & Viswanathan, V. (2018). Performance metrics required of next-generation batteries to electrify vertical takeoff and landing (VTOL) aircraft. *ACS Energy Letters, 3*(12), 2989–2994.

Garrow, L. A., German, B. J., Schwab, N. T., Patterson, M. D., Mendonça, N. L., Gawdiak, Y. O., & Murphy, J. R. (2022). A proposed taxonomy for Advanced Air Mobility. *Proceedings of the AIAA Aviation Forum 2022.*

Goldstein, B. (2024, April 8). *EHang obtains production certificate for EH216-S from CAAC.* Aviation Week Network. https://aviationweek.com/aerospace/advanced-air-mobility/ehang-obtains-production-certificate-eh216-s-caac

Guo, J., Chen, L., Li, L., Na, X., Vlacic, L., & Wang, F. (2024). Exploring the economic feasibility of Advanced Air Mobility in the early stages. *IEEE Transactions on Intelligent Vehicles, 9*, 4826–4830. https://doi.org/10.1109/TIV.2024.3422002

Hawkins, A. J. (2025, April 17). *Archer and United plan NYC air taxi service for airport trips. The Verge.*

Holden, J., & Goel, N. (2016, October). Fast-forwarding to a future of on-demand urban air transportation (Uber Elevate White Paper). *Uber Technologies.* https://evtol.news/__media/PDFs/UberElevateWhitePaperOct2016.pdf

Holder, B., & Mitchell, L. (2023, January 23). *NASA goes live with surrogate eVTOL flight tests in Texas.* NASA.

Joby Aviation. (2023a, June 28). *Joby marks production launch, receives permit to fly first aircraft built on production line* [Press release]. https://ir.jobyaviation.com/news-events/press-releases/detail/64/joby-marks-production-launch-receives-permit-to-fly-first

Joby Aviation. (2023b, September 25). *Joby delivers first eVTOL aircraft to Edwards Air Force Base ahead of schedule* [Press release]. https://www.jobyaviation.com/news/joby-delivers-first-evtol-edwards/

Joby Aviation. (2024, February). Joby to launch air taxi service in UAE (RTA/Skyports agreement). *Joby Aviation, Inc.*

Joby Aviation. (2025, June 30). *Joby cements global lead in air taxi industry with Dubai flights and beginning of commercial market readiness work.* Joby Aviation.

Khavarian-Garmsir, A., Sharifi, A., & Sadeghi, A. (2023). The 15-minute city: Urban planning and design efforts toward creating sustainable neighborhoods. *Cities, 138*, 104101. https://doi.org/10.1016/j.cities.2022.104101

Molyneaux, I. (2023, June 29). *FAA allows first production Joby eVTOL aircraft to start test flights. AeroTime.* https://www.aerotime.aero/articles/joby-aviation-evtol-faa-special-airworthiness-certificate

NASA. (2005). The Bell XV-15 tilt-rotor research aircraft: From concept to flight (Monographs in Aerospace History No. 17). *NASA History Division.*

NASA. (2019). eVTOL workshop final report (NTRS 20190028296). *NASA Technical Reports Server.* https://ntrs.nasa.gov/api/citations/20190028296/downloads/20190028296.pdf

NASA. (2020a). NASA's Urban Air Mobility Grand Challenge advances with agreement signings (Release 20-018). *NASA.* https://www.nasa.gov/news-release/nasas-urban-air-mobility-grand-challenge-advances-with-agreement-signings/

NASA. (2020b). Urban Air Mobility: Grand Challenge—Goals & objectives (NTRS 20200001287). *NASA Technical Reports Server.* https://ntrs.nasa.gov/api/citations/20200001287/downloads/20200001287.pdf

National Aeronautics and Space Administration (NASA). (n.d.). Advanced Air Mobility mission. *NASA Aeronautics Mission* page. Retrieved July 4–5, 2025, from https://www.nasa.gov/mission/aam/

National Museum of the U.S. Air Force. (n.d.). Avro Canada VZ-9 Avrocar. *National Museum of the U.S. Air Force.*

National Transportation Safety Board (NTSB). (1977). Aircraft Accident Report, New York Airways S-61L, N619PA, Pan Am Building Heliport, New York, May 16, 1977 (AAR-77-09). *National Transportation Safety Board.* https://www.ntsb.gov/investigations/AccidentReports/Reports/AAR7709.pdf

Polmar, N. (2013a). *Historic Aircraft – The Tail-sitters, Part I—Lockheed. Naval History Magazine* (U.S. Naval Institute).

Polmar, N. (2013b). *Historic Aircraft – The Tail-sitters, Part II—Convair. Naval History Magazine* (U.S. Naval Institute).

Pozoukidou, G., & Chatziyiannaki, Z. (2021). 15-Minute city: Decomposing the new urban planning eutopia. *Sustainability, 13*(2), 928. https://doi.org/10.3390/su13020928

Reuters. (2024, October 22). *Hyundai's Supernal partners with two firms for its battery-powered aircraft.* Reuters. https://www.reuters.com/business/autos-transportation/hyundais-supernal-partners-with-two-firms-its-battery-powered-aircraft-2024-10-22/

Ridden, P. (2024, August 9). *Stunning eVTOL lifts off after setback of Olympic proportions.* New Atlas.

Roads and Transport Authority (RTA) Dubai. (2024). Al Masar 189: Aerial taxis in Dubai by 2026 (RTA–Skyports–Joby roles). *RTA.*

Ros, M. (2025, June 30). *Joby performs piloted full-transition test flights in Dubai.* AeroTime.

Science Museum Group (London). (n.d.). Rolls-Royce Thrust Measuring Rig ("Flying Bedstead")—First jet-lift VTOL rig. *Science Museum Blog.*

SFGATE (Glab, J.). (2025, June 18). This helicopter airline connected the Bay Area to SFO in minutes; SFO Helicopter Airlines history (travel feature). *SFGATE.*

SkyDrive, Inc. (2025, July 31). Starts summer demonstration flights at Expo 2025 Osaka. *SkyDrive* (en.skydrive2020.com).

Smithsonian National Air and Space Museum. (n.d.-a). Emile and Henry Berliner—Early helicopter experiments. *Smithsonian NASM.*

Smithsonian National Air and Space Museum. (n.d.-b). Rocket Belt, Bell No. 2 (collection page). https://www.airandspace.si.edu/collection-objects/rocket-belt-bell-no-2/nasm_A19730264000

Sripad, S., & Viswanathan, V. (2021). The promise of energy-efficient battery-powered urban aircraft. *Proceedings of the National Academy of Sciences, 118*(45), e2111164118. https://www.pnas.org/doi/pdf/10.1073/pnas.2111164118

Stonor, C. (2025, July 3). *Archer begins trial flights in Abu Dhabi. eVTOL Insights.*

Tang, H., Zhang, Y., Mohmoodian, V., & Charkhgard, H. (2021). Automated flight planning of high-density urban air mobility. *Transportation Research Part C: Emerging Technologies, 131,* 103324. https://doi.org/10.1016/j.trc.2021.103324

The Editors of Encyclopaedia Britannica. (n.d.-a). Paul Cornu. In *Encyclopaedia Britannica.* Retrieved August 9, 2025, from https://www.britannica.com/biography/Paul-Cornu

The Editors of Encyclopaedia Britannica. (n.d.-b). Fw 61. In *Encyclopaedia Britannica.*

The Editors of Encyclopaedia Britannica. (n.d.-c). Harrier. In *Encyclopaedia Britannica.*

TOI World Desk. (2025, June 14). *Driverless flying taxi in UAE soon? Watch Abu Dhabi conduct test flight over Marina. The Times of India.*

U.S. Air Force/AFWERX (via *Avionics International*). (2022, January–February). The U.S. Air Force Agility Prime program: Progress in 2021 and goals for 2022. https://interactive.aviationtoday.com/avionicsmagazine/january-february-2022/the-u-s-air-force-agility-prime-program-progress-in-2021-and-goals-for-2022/

Varghese, J. (2025, October 27). Drone delivery takes flight in Nad Al Sheba as Dubai launches smart, sustainable route. *Gulf News.* Retrieved from https://gulfnews.com/business/markets/drone-delivery-takes-flight-in-nad-al-sheba-as-dubai-launches-smart-sustainable-route-1.500323073

Vidalon, D. (2024, August 8). *Plans to fly air taxis with passengers during Paris Olympics scrapped.* Reuters. (Reprinted via MarketScreener).

Volocopter. (2019, October 22). *Volocopter air taxi flies over Singapore's Marina Bay* [Press release].

Volocopter. (2021, November 11). *Volocopter conducts South Korea's first crewed public air taxi test flight* [Press release].

Volocopter/Dubai RTA. (2017, September 25). First-ever public demonstration of an autonomous urban air taxi (Dubai). *Volocopter newsroom; RTA Al Masar* (issue 113).

Whiting, T. (2022, May 10). *Science of Sound: NASA examines advanced air mobility noise.* NASA.

wow-RAK. (2025, April 22). *Ras Al Khaimah to launch the region's first electric flying car service.* wow-RAK. https://wow-rak.com/ras-al-khaimah-electric-flying-car/

Xie, Y., Pongsakornsathien, N., Gardi, A., & Sabatini, R. (2021). Explanation of machine-learning solutions in air-traffic management. *Aerospace*, *8*(8), 224. https://doi.org/10.3390/aerospace8080224

2 AAM—Concentric Layers

INTRODUCTION

Implementing advanced air mobility (AAM) is a multidisciplinary challenge. It can be thought of as an ecosystem with several interdependent components, each of which must develop in harmony with the others for the overall system to succeed. The five key layers (Figure 2.1)—aircraft, infrastructure, airspace, operations, and community—are widely recognized as foundational to the development and deployment of AAM systems. These five—overlapping and interdependent—layers encompass the vehicles that will fly, the ground/support infrastructure they require, the airspace system that will manage them, the operational frameworks for their use, and the human/social dimension of public acceptance. In short, AAM is not just about inventing a new aircraft or app—it is about orchestrating a whole new mode of transportation through a system-of-systems approach. Each layer must progress in concert with the others, while addressing each layer holistically is essential for building a safe, efficient, and publicly supported AAM ecosystem.

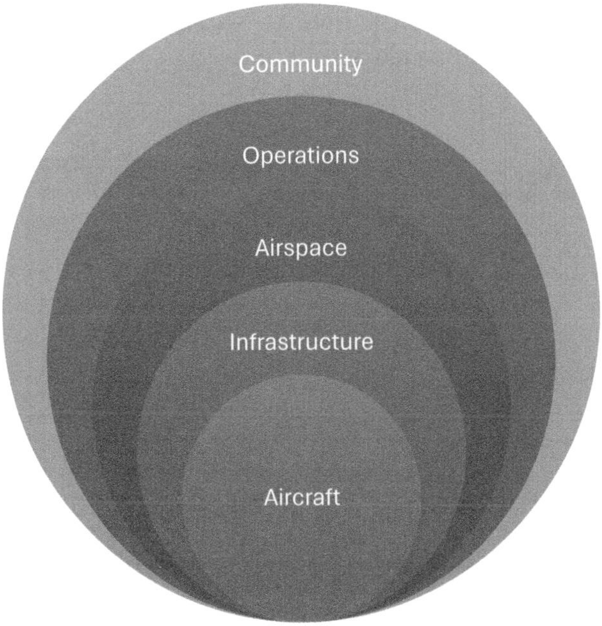

FIGURE 2.1 AAM's Concentric layers. (Credit: Author.)

DOI: 10.1201/9781003546306-3

AIRCRAFT

AAM aircraft—principally electric vertical takeoff and landing (eVTOL) vehicles, but also short takeoff and landing (STOL) concepts and advanced rotorcraft—provide the technological core of the AAM vision. These vehicles are optimized for short-haul, low-altitude flights and seek to combine high safety standards with operational efficiency and improved environmental performance (Johnson & Silva, 2021; Kiesewetter et al., 2023). The manufacturing sector has evolved from a handful of pioneers in 2010 to a rapidly expanding global industry, with forecasts of strong growth in the number of companies and market value by 2040 (see Figure 2.2).

Contemporary AAM designs cluster around a small set of configurations whose trade-offs are now well documented in the technical literature. *Lift-plus-cruise* architectures use dedicated vertical-lift propulsors alongside separate cruise propellers; *tilt-wing* and *tiltrotor* concepts reorient lifting surfaces or rotors to transition between hover and wing-borne flight; *multirotor* concepts scale up drone-like arrangements for people or cargo; and *advanced rotorcraft* evolve conventional helicopters for lower noise and higher efficiency (Johnson & Silva, 2021).

Propulsion spans battery-electric, hybrid-electric, and turboelectric options (see Figure 2.3 and Table 2.1). Most early vehicles carry roughly one to eight passengers or equivalent payloads, with near-term battery-electric ranges commonly on the order of ~20–100 km per charge for urban missions, while some hybrid/turboelectric concepts—especially larger tilt-wing/tiltrotor studies—target regional missions of hundreds of nautical miles (Johnson & Silva, 2021; Kiesewetter et al., 2023). Notably, NASA's reference fleet has included a 15-passenger tilt-wing with turboelectric propulsion sized for ~400 nm, illustrating the step-change possible when moving beyond pure battery-electric architectures (Guo et al., 2024; Johnson & Silva, 2021; Silva et al., 2018).

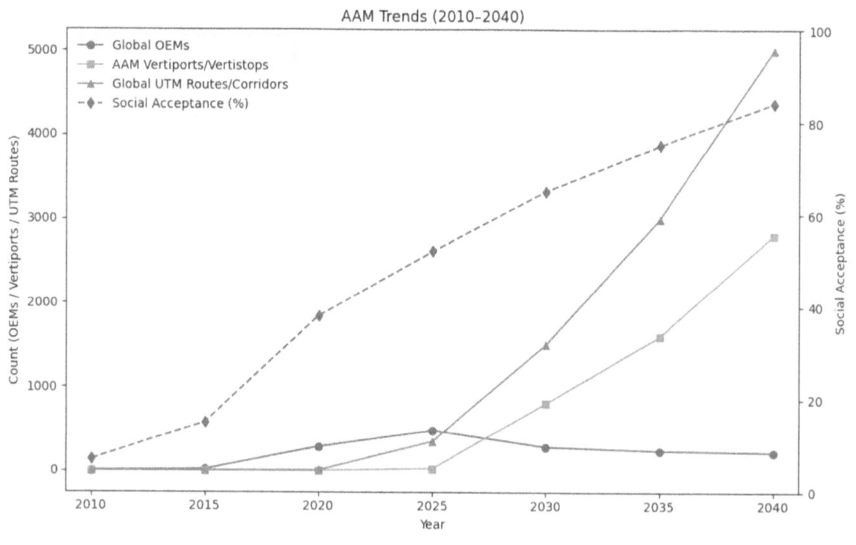

FIGURE 2.2 Chronological evolution of AAM's layers.[1] (Credit: Author.)

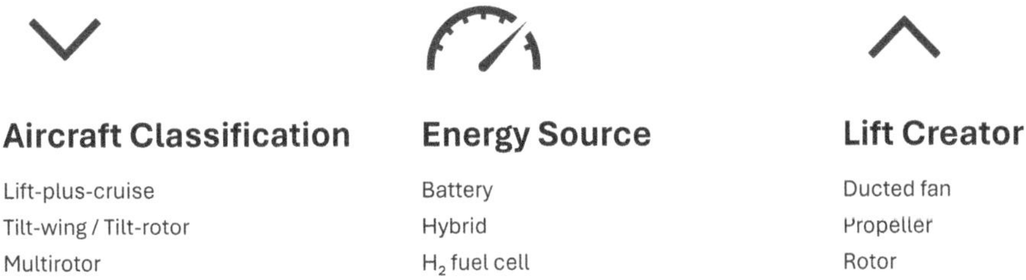

FIGURE 2.3 AAM aircraft and propulsion classification. (Credit: Author.)

CURRENT DEVELOPMENTS

AAM aircraft development is progressing quickly, with dozens of eVTOL and related designs moving through concept definition, flight testing, and regulatory engagement (over 800 concepts are listed in the Vertical Flight Society's (2025) website). As of mid-2025, no passenger-carrying eVTOL has received FAA or EASA type certification; frontrunners such as Joby Aviation (Stubbins, 2025) report advancing through

TABLE 2.1
AAM Aircraft and Propulsion Classification

Aircraft Classification	Energy Source	Propulsion Approach	Representative Propulsor(s)
Lift-plus-cruise	Battery-electric	Separate hover + cruise electric systems	Multiple small lift fans + wing cruise fan/prop
	Hybrid-electric	Turbogenerator or combustion + generator + batteries	Lift fans + wing prop driven electrically
	Turboelectric/ H_2-combustion	Gas turbine → generator → electric motors (can burn H_2)	Distributed fans/props
Tilt-wing/ Tiltrotor	Battery-electric	All-electric tilt mechanisms + motors	Tiltrotors/tilting wing-mounted props
	Hybrid-electric	Hybrid turbogenerator + batteries	Tiltrotors + electric drive
	Turboelectric/H_2	Large turbogenerator/turboelectric or H_2-turbine → generator	Large tiltrotors, distributed electric fans
Multirotor (distributed lift)	Battery-electric	Pure electric distributed lift	Many small rotors/ducted fans
	Hybrid-electric/ Turboelectric	Hybridized generator + batteries	Distributed rotors with hybrid power
Advanced rotorcraft (evolved helicopter)	Battery-electric	Electric main rotor (partial) + batteries	Large main rotor/ anti-torque systems
	Hybrid-electric	Turboshaft + electrical augmentation/turbogenerator	Main rotor driven via hybrid drivetrain
	Turboshaft/ Turboprop	Conventional turbine engines	Main rotor/propeller

Type Inspection Authorization activities toward U.S. certification, and recent disclosures point to 2026-onward service entries contingent on final approvals (Reuters, 2025). In China, by contrast, the Civil Aviation Administration of China (CAAC) has issued a type certificate (October 2023), a production certificate (April 2024), and standard airworthiness certificates (December 2024) for EHang's EH216-S—enabling supervised public demonstrations and early commercial deployments under Chinese oversight (EHang Holdings Limited, 2023a, 2023b, 2024).

AAM vehicle architectures span lift-plus-cruise, tilt-wing, tiltrotor, multirotor, and advanced rotorcraft configurations. Across these families, the design task is to balance vertical performance for takeoff/landing with efficient wing-borne cruise, while minimizing exterior noise and meeting airworthiness targets. Propulsion choices range from battery-electric to hybrid- or turboelectric systems; near-term battery-electric vehicles are generally optimized for short urban or peri-urban sectors, whereas some hybrid/turboelectric concepts target longer regional missions (Guo et al., 2024; Johnson & Silva, 2021; Kiesewetter et al., 2023). NASA's concept studies and engineering analyses synthesize these trade-offs and have been used in investigations of handling qualities and community noise below helicopter baselines (Johnson & Silva, 2021).

Testing and certification programs increasingly blend flight testing with high-fidelity simulation. Digital twin methods and multi-agent simulation environments are being used to prototype flight-control behaviors, evaluate human-in-the-loop interactions, and rehearse airspace scenarios before costly real-world trials (Conrad et al., 2023; Flores et al., 2023; Kannan et al., 2023; Ziakkas et al., 2024). In parallel, NASA's AAM National Campaign has executed surrogate eVTOL operations and full-scale acoustic campaigns to provide regulators and communities with measured data across representative mission profiles—work that feeds directly into noise, integration, and safety assessments (NASA, 2022a, 2022b).

From a regulatory standpoint, U.S. and European authorities remain closely engaged with manufacturers. Public communications emphasize a safety-first trajectory—piloted initial operations, progressive automation, and iterative guidance as evidence accumulates. Industry scans consistently describe dozens of programs pursuing certification in the United States and Europe, a count that has fluctuated as programs consolidate or pivot (L.E.K. Consulting, 2021). Against this backdrop, leading prototypes continue to log flight hours, expand to new test sites, and announce infrastructure partnerships while final compliance testing proceeds (AeroTime, 2025; Reuters, 2025).

Development typically progresses through three overlapping stages. First, in concept and design, teams select configurations and propulsion architectures and iterate rapidly using simulation and digital twins to explore performance, controllability, maintenance, and acoustic signatures (Conrad et al., 2023; Johnson & Silva, 2021; Kiesewetter et al., 2023). Next, in prototype and testing, they expand flight envelopes, conduct human-factors evaluations, and validate safety cases against expected operations (Flores et al., 2023; Kannan et al., 2023; Ziakkas et al., 2024). Finally, during certification, manufacturers complete conformity builds, submit artifacts, and execute regulator-witnessed tests toward type and production approvals, followed by operational authorization (Johnson & Silva, 2021; Kiesewetter et al., 2023).

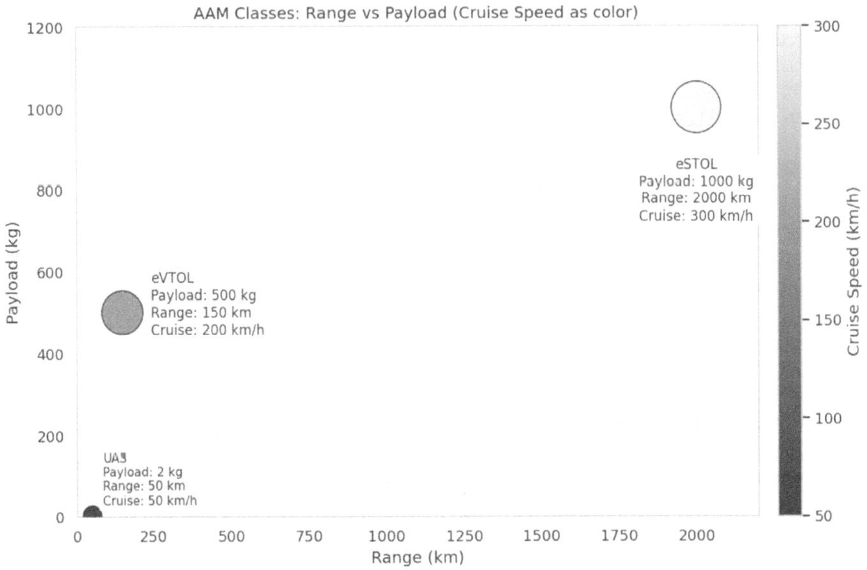

FIGURE 2.4 AAM's generic aircraft performance. (Credit: Author.)

While the exact sequencing varies by authority and applicant, this path now has enough definition that both industry and government can align roadmaps and readiness criteria.

Figure 2.4 illustrates a generic level of performance for AAM's aircraft in reference to payload, range, and cruise speed. The Unmanned Air Systems (UASs), based on their performance, are used for last-mile delivery at low altitude, short range, and minimal payload. Then the eVTOLs exhibit limited payload and range, largely due to lithium-ion battery constraints, while the electric STOLs (eSTOLs) deliver regional transport capabilities at higher speed, with the use of hybrid systems enabling high range and payload. Other characteristics are also essential, such as the level of noise, weather limitations, and maintenance/turnaround needs, which differentiate each class in their scope and potential areas of usage.

INFRASTRUCTURE

The Infrastructure layer covers the physical and organizational facilities needed on the ground (and in the network) to support AAM. Primarily, this means vertiports or vertistops—the equivalent of airports for eVTOLs—where vehicles can take off and land, recharge or refuel, load and unload passengers or cargo, and be maintained. Infrastructure also encompasses the electrical power supply for charging, the communications and navigation installations (like ground antennas or data centers for managing AAM traffic), and integration with existing transport infrastructure (roads, public transit, airports). In broader terms, one can include policy and standards infrastructure here as well (though those could be seen as part of Community/

Governance). Essentially, without adequate infrastructure, AAM vehicles will have nowhere to fly and no way to operate at scale.

Current Developments

Recognizing the critical role of vertiports, aviation authorities and standards bodies have begun issuing guidelines to ensure safe and efficient vertiport design related to the following developments:

Regulatory Guidance

EASA published a Prototype Technical Design Specifications for Vertiports in 2022 (EASA, 2022). This document provides a template for how vertiports should be laid out and equipped. It covers elements like: the touchdown and liftoff area where the eVTOL actually lands, the final approach and takeoff paths that aircraft use to approach/depart, obstacle clearance surfaces around the vertiport (to ensure no tall object intrudes into the flight path), safety buffers and fencing, lighting and markings (e.g., a big "V" or similar marking might denote a vertiport pad, analogous to the "H" for heliports), firefighting and rescue equipment on site, and provisions for charging infrastructure. Similarly, the FAA (2025) released Engineering Brief No. 105 in 2022 (updated as 105A in 2025), which gives interim standards for vertiport design in the United States. The FAA's guidance classifies vertiports by the size/class of eVTOL they serve and provides dimensions for things like pad size and separation based on the rotor diameter of the aircraft (using a concept of "RD"—rotor diameter—to size landing areas). It also adds features such as a downwash/outwash protection zone to account for the high-speed airflow generated by multiple rotors. These standards are evolving as data from testing comes in, but they serve as a starting point so that infrastructure planners know what requirements to meet. By standardizing key parameters (e.g., weight support, charging interfaces), they ensure that any compliant eVTOL can land at any compliant vertiport, which will be important for interoperability.

Vertiport Types and Locations

We can categorize vertiport infrastructure by size and role. A *vertiport* generally implies a facility with multiple pads and perhaps amenities for passengers (waiting lounge, etc.), whereas a *vertistop* might be a single-pad, minimal facility meant for quick pickups/drop-offs (like a taxi stand). In cities, we might see a few larger vertiports (hubs) and many smaller vertistops. Likely locations include: existing airports (many airports are planning eVTOL facilities on site to link to downtown), rooftops of large buildings (helipads on hospitals, hotels, or parking structures that can be upgraded to vertiports), waterfront areas (using barge or pier installations), and ground-level sites like repurposed parking lots. Real estate considerations are big-city centers have limited open space, so rooftops are attractive.

In 2022, a U.K. company opened a demonstrator called *Air-One* in Coventry, a pop-up vertiport on the ground, to showcase AAM infrastructure; it even had a drone takeoff and landing to simulate operations (Arup, 2022). Such demos are being used to figure out issues like crowd flow and noise management. Integration with existing transport is key: ideally, vertiports are co-located with other hubs—e.g., at a train station or major transit stop—so that passengers can easily transfer from ground transport to air. Studies emphasize that vertiport placement should leverage the existing public transit network, prioritizing highly connected nodes such as major subway or metro stations to maximize accessibility and network efficiency (Rahman et al., 2023; Song & Lee, 2025; Zhao & Feng, 2024). In Seoul and Beijing, methodologies using network centrality and clustering algorithms have been applied to identify optimal vertiport locations, often aligning with major transit hubs to ensure robust connectivity and significant travel time savings (up to 80%) (Lim & Hwang, 2019; Song & Lee, 2025; Zhao & Feng, 2024), since integrating vertiports with multimodal hubs (e.g., metro, bus, taxi) is shown to improve overall transportation network performance, reduce travel distances, and enhance user convenience. Planners also consider equity: if vertiports only serve wealthy downtown districts, the benefits of AAM won't be widespread. Some cities are examining placing vertiports in underserved areas or at university campuses, etc., to distribute access.

Power and Energy Infrastructure

Since most AAM aircraft are expected to be electric, the power supply is arguably the most critical infrastructure element aside from the landing pad itself. High-voltage charging stations will be installed at vertiports to recharge eVTOL batteries between flights. The power requirements can be substantial: Fast-charging eVTOLs to enable quick turnarounds (5–10 minutes) can require power inputs from several hundred kilowatts up to 1 megawatt, far exceeding the 150–250 kW of most current EV fast chargers (Hays et al., 2024; He et al., 2024; Yang et al., 2021). If multiple aircraft need to charge concurrently, the vertiport effectively becomes a small urban electric substation in terms of load (see Figure 2.5). This raises challenges: grid connectivity—can the local grid feed that much power, especially during peak times?

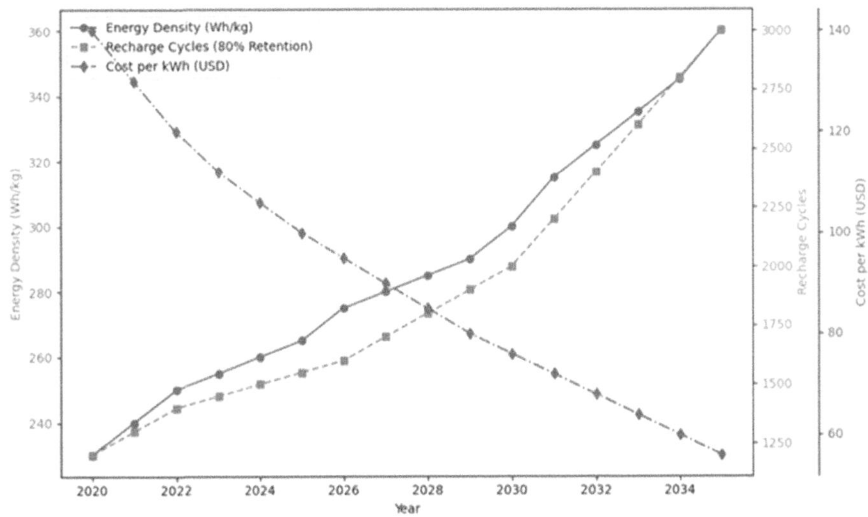

FIGURE 2.5 Battery and energy storage trends and projections (2020–2035). (BloombergNEF, 2023; Cole & Karmakar, 2023; International Energy Agency, 2023; Liu, 2023)[2]

Utilities may need to upgrade lines or transformers leading to vertiport sites. Some vertiports might incorporate on-site energy storage (large battery banks or possibly stationary flywheels) to buffer the load—they could slowly draw power from the grid continuously, then dump it quickly into vehicles when needed. Others might explore solar panels or other renewables on-site, though urban locations often have limited space for that, and solar is intermittent.

Standardization of charging connectors will be important so that any eVTOL can use any vertiport's charger (similar to how Combined Charging System (CCS))[3] is a standard for electric cars, albeit Tesla had its own). Efforts by SAE International are underway to define an aeronautical electric charging standard, possibly involving automated charging connections (like robotic arms or pad induction).

Another approach some manufacturers consider is battery swapping: instead of charging the aircraft on the spot, a depleted battery module could be quickly swapped out for a charged one. This can reduce turnaround time (a swap could be a few minutes versus 20–30 minutes to charge). However, swapping requires a stock of spare batteries on site and machinery to do the swap, and it ties the operation to a particular battery design standard (hard to do across multiple manufacturers). Only a few eVTOL designs (like EHang's two-seater) have explicitly touted swapping.

Hydrogen fuel infrastructure could come into play when hydrogen fuel cell eVTOLs are certified for broader use[4]—that introduces the need for hydrogen storage (likely as compressed gas or liquid) and refueling apparatus, which has its own safety (risks such as leaks, embrittlement, and potential ignition) and regulatory considerations.

Maintenance and Support Facilities

Like any aircraft, eVTOLs will require regular maintenance—some tasks can be quick turnarounds (a daily check, swap of a modular component) while others might require taking the aircraft out of service for hours or days. Infrastructure must accommodate maintenance needs either at vertiports or at separate maintenance bases. Some vertiports (especially larger ones or those at airports) may include a hangar or workshop area where technicians can perform routine inspections, repairs, and software updates. Other networks might designate one central maintenance base in a region and ferry aircraft there as needed (this is easier if that base is within ferry range and there's enough slack in the schedule to pull a vehicle out).

Vertiport design should consider maintenance logistics, for example, having a safe area to tow a vehicle to a hangar, an area for spare parts storage, etc. On the software side, operators will use health monitoring systems on the aircraft (sensors that track motor health, battery health, etc.) and integrate that with operations scheduling—if a sensor flags a potential issue, the operations control might automatically route that aircraft to a maintenance site after its current flight and dispatch a backup to cover its next scheduled trip. This predictive maintenance approach can minimize unplanned downtime. But it requires that the maintenance infrastructure (tools, trained staff, and possibly remote diagnostic support from the manufacturer) is in place to quickly address issues.

Another aspect is charging infrastructure doubling as maintenance: some battery issues might be detected at charging stations (battery management systems communicating with the charger), so vertiports will also be data hubs, sending vehicle health info to the cloud for analysis.

AIRSPACE

The Airspace layer of AAM refers to the air traffic management (ATM) frameworks, rules, and systems needed to safely accommodate a new class of vehicles in the skies. Even the most advanced eVTOL is of little use if it cannot fly in a coordinated way without conflict with other aircraft or violation of regulations. Traditional aviation has a well-established ATM system for commercial airliners and general aviation, operating under instrument or visual flight rules. AAM, however, envisions potentially hundreds or thousands of aircraft operating at low altitudes (generally below a few thousand feet) in urban and suburban areas—airspace that current ATM systems and procedures are not designed to handle at such a scale. Thus, the airspace layer is about developing new approaches to separation assurance, traffic flow management, and integration. This includes concepts like UAS Traffic Management (UTM) adapted for urban air mobility, designated corridors or routes for eVTOL operations, autonomous flight rules, new communication/navigation requirements, and possibly greater automation in air traffic services.

SIGNIFICANCE

Effective airspace integration is critical for safety—the top priority in aviation. Without proper airspace management, the risk of mid-air collisions or interference with existing aircraft would be unacceptable, which could halt AAM before it starts. Additionally, the capacity and scalability of AAM hinge on airspace system capabilities. The difference between handling a dozen air taxis in a city versus scaling to hundreds or more is largely an airspace coordination problem (assuming the vehicles themselves are available).

NASA introduced the concept of UAM maturity levels (UML) to describe increasing levels of airspace and operational complexity; at higher UML (e.g., UML-4 and UML-5), it anticipates *medium to high density and complexity operations with collaborative and automated systems* managing the traffic (NASA ARMD, 2021). Achieving those levels will require sophisticated automation in traffic management and new roles for both humans and machines in controlling air traffic.

Airspace design for AAM also has a strong influence on community acceptance and policy: for instance, establishing designated routes or "sky lanes" could help concentrate and manage noise away from sensitive areas, and setting altitude floors/ceilings can alleviate privacy concerns on the ground. On the other hand, overly rigid corridors might concentrate noise and raise equity issues (e.g., which neighborhoods are under the flyways?). These considerations show how tightly the Airspace layer links to the Community layer as well as Operations.

Current Concepts and Initiatives

Several efforts are underway to define how AAM will navigate the skies:

FAA's UAM Concept of Operations (ConOps)

The FAA published a UAM ConOps v1.0 in 2020 and an updated ConOps v2.0 in 2023. These documents lay out an evolving strategy for integrating early AAM operations into the National Airspace System. The vision in ConOps 2.0 (FAA, 2023) is that initial AAM services will use incremental integration: in early stages, eVTOL flights might operate much like helicopter flights today (with pilots onboard, talking to air traffic control in certain airspace, following published routes or corridors). The ConOps suggests use of *prioritization corridors* at low altitudes where AAM traffic could be funneled, possibly connecting key nodes like airports and downtown vertiports. Over time, as traffic density increases, the system would introduce more automation and network-based deconfliction. The FAA is leveraging experiences from drone integration—for example, requiring AAM operators to have equipment like ADS-B Out or transponders in certain airspace (just as other aircraft do) so they are visible on ATC radar (FAA, 2023). A notable point is that *near term AAM flights will likely be handled within existing airspace classes and rules*. This means an eVTOL flying in controlled airspace must follow similar communication and surveillance requirements as a helicopter, which ensures safety but may constrain operations until new rules are developed. ConOps v2.0 emphasizes a *cooperative traffic management* approach, where vehicles, service providers, and ATC share data to maintain separation with less micromanagement by human controllers. This cooperative approach is akin to how UTM for drones works (with digital sharing of flight intent and automated conflict advisories).

UAS Traffic Management Evolution

NASA and others have been developing UTM for small unmanned aircraft, and these concepts are being extended to AAM. UTM relies on a distributed, highly automated system in which approved service providers help manage air traffic through information exchange rather than centralized control. For AAM, one can envision a system where flight plans are submitted to a cloud-based coordination service that deconflicts trajectories in advance (strategic deconfliction) and provides real-time advisories to aircraft if conflicts arise (tactical deconfliction). Researchers have proposed that above certain densities of operation, conventional *see-and-avoid* or manual air traffic control will not be sufficient—instead, vehicles will need to have detect-and-avoid systems and be linked into a network that coordinates separation automatically (Thipphavong et al., 2018). For example, a network of rooftop "urban radar" or surveillance nodes might track AAM flights, and if two flight paths are predicted to conflict, the system could automatically suggest or impose adjustments to their altitudes or routes.

NASA's airspace integration concepts (Thipphavong et al., 2018) describe a future where third-party UAM service suppliers assist in managing flows, similar to how internet service providers route data. In practice, this might mean that an AAM operator is subscribed to a traffic management service that handles its flight clearances and sequencing with other operators, only involving traditional ATC when interfacing with airports or higher-altitude airspace. Several field trials of UTM for drones have already been conducted (in the United States, Europe, and elsewhere), and these are informing how a UTM-like system for AAM could function.

U-space in Europe

In Europe, the U-space framework is the counterpart to UTM, and it reached a milestone in January 2023 when new EU regulations for U-space airspaces came into effect. U-space designates certain volumes of airspace (likely in low-altitude urban areas) where any unmanned or advanced aircraft must use digital services for traffic management. These services include network identification, geo-fencing, flight authorization, and real-time traffic information. For future air taxis, integrating into U-space means equipping the vehicles to communicate with these services—essentially, each flight would be tracked and managed via networked systems rather than solely by human observers.

The European approach is a phased implementation: member states are now tasked with designating U-space airspace and providers, and the system is expected to mature through the 2020s, enabling more complex operations (European Commission, 2023).

Low Altitude Airspace in China

China's *low-altitude economy* is moving from policy concept to implementation, and nowhere more visibly than in Shenzhen, where municipal authorities are piloting digital airspace services, expanding corridors, and codifying responsibilities for planning and operations. At the national and city levels, the technical thrust centers on digitizing low-altitude airspace, defining rights and responsibilities in three dimensions, and enabling high-density UAV/AAM flows in complex urban canyons. One strand formalizes low-altitude space within land administration: using the Land Administration Domain Model to delineate urban low-altitude traffic space, identify fuzzy boundaries, and confront cross-departmental coordination—work explicitly demonstrated through Shenzhen case studies (Li et al., 2024). In parallel, Shenzhen has adopted enabling policies and targets—including plans for more than 1,000 takeoff/landing sites and more than1,000 commercial routes by end of 2025 and new municipal regulations granting authority to plan and operate low-altitude infrastructure—signaling that airspace services, infrastructure, and regulations are being advanced together rather than in isolation (Shenzhen Daily, 2024a, 2024b; Zhang, 2024).

The technical backbone for high-density operations is a digital twin of the urban airspace with real-time 3D geospatial "mining," airspace discretization, and trajectory optimization—an approach shown to increase capacity and maintain compliance by embedding policy constraints into the routing fabric (Elsayed & Mohamed, 2024; Namuduri, 2023). Recent studies also point to multi-agent reinforcement learning for structured AAM traffic, where agents co-optimize separation, delay, and energy in constrained corridors—methods directly relevant to managing Shenzhen's anticipated flows (Deniz et al., 2024). Nevertheless, scaling from pilots to citywide service still hinges on communications coverage, latency, and data-exchange reliability for command-and-control and strategic deconfliction—concerns highlighted in AAM systems work that couples operations and infrastructure with network performance (Al-Rubaye et al., 2023).

Within this maturing airspace fabric, Meituan illustrates how commercial operators may leverage policy and technology advances. In April 2025, Meituan obtained the country's first nationwide low-altitude logistics operating certificate from the CAAC, enabling routine BVLOS drone logistics across multiple cities (including Shenzhen) under standardized oversight; the milestone reflects the regulator's readiness to normalize recurrent, digitally supervised operations rather than one-off demos (*China Daily*, 2025; TechNode, 2025a). The broader ecosystem around Shenzhen continues to layer services on top of this foundation—for example, 3D low-altitude navigation products that fuse mapping, route planning, and collision-warning functions for public and enterprise users—hinting at a future where digital services and airspace rules co-evolve (TechNode, 2025b).

Communication and Navigation Infrastructure

The Airspace layer also involves ensuring that AAM vehicles have reliable communication links and navigation capabilities in the environments they will operate. Many AAM operations will occur at low altitude in urban areas where traditional aviation infrastructure is sparse: radar coverage may be limited below building heights, and ground-based navigation aids (like VOR/DME[5]) are irrelevant to these aircraft, which will rely on GPS and inertial systems. Therefore, dependence on satellite navigation and augmentations will be high, raising concerns about vulnerabilities like GPS jamming or multipath interference in cities. As backups, some propose using signals like enhanced LORAN or leveraging 5G cellular networks for positioning.

Indeed, it has been suggested extending 5G/6G telecommunications infrastructure to support AAM communication and navigation needs, given 5G's low-latency and high-bandwidth capabilities (Arafat & Pan, 2024). This hints at the crossover between airspace and infrastructure layers: digital infrastructure (communication antennas, data centers, spectrum allocation) becomes as crucial as physical infrastructure. For example, if an eVTOL depends on a constant data link for traffic information or remote pilot control, cities might need to install ground stations or mesh networks to ensure coverage along all AAM routes. Navigation in urban canyons (among tall buildings) is another challenge—multi-sensor integration (GPS, vision-based navigation, etc.) is being developed to keep vehicles on track even when one system falters. Robust surveillance systems will also be needed, such as networks of ADS-B receivers or drone-detect radars, to monitor AAM traffic for authorities and provide a common picture of the low-altitude airspace.

OPERATIONS

The Operations layer of AAM encompasses how AAM services will be run on a day-to-day basis. This includes the operational concepts (ConOps) for different AAM missions, the business models and use cases (e.g., air taxi for commuters, cargo delivery, emergency response, etc.), the procedures for flight operations (from scheduling and dispatch to in-flight management and ground handling), and the regulatory and organizational structures for operators. In traditional aviation, *operations* might refer to an airline's flight operations department or a helicopter charter company's procedures. In AAM, operations could take new forms—for instance, a ride-sharing model where flights don't have fixed schedules but respond to on-demand requests via an app, or autonomous cargo networks running frequent short hops. Essentially, this layer addresses how AAM flights are actually conducted and monetized.

Significance

The operations layer is where the value proposition of AAM is realized. It translates the existence of aircraft and airspace procedures into tangible mobility services that customers use. Well-designed operations ensure efficiency, safety in practice, and a positive user experience. Poorly designed operations, by contrast, could lead to underutilization (e.g., empty or idle aircraft much of the time), high costs, or even safety incidents—even if the vehicles and airspace system are technically sound.

In academic terms, operations research and systems engineering play a big role in this layer: optimizing routes, schedules, fleet sizes, maintenance intervals, and pricing strategies are all operational considerations. For AAM to fulfill promises like reduced travel time or lower emissions per trip, the operations must be tightly managed. This means minimizing turnaround times at vertiports, efficiently recharging or swapping batteries, and smartly repositioning vehicles to meet demand (similar to how rideshare companies reposition cars to serve hot spots).

Key Aspects

Autonomy and Human Factors

A major operational evolution will be the shift from piloted to highly automated or autonomous operations. This raises questions of *how* operations are conducted and monitored. In early years, if every eVTOL has a pilot on board, operations will resemble today's helicopter ops, but potentially at a greater scale (many more flights per hour, which means many more pilots needed). Over time, companies plan to introduce greater autonomy. For instance, Wisk Aero's (n.d.) design is for a fully autonomous air taxi from the start (no pilot, just remote supervision), though certifying that is challenging.

Operationally, transitioning to autonomy means creating new roles: the concept of a "fleet supervisor" or "mission manager" on the ground who monitors multiple flights is often discussed. This is analogous to how operators manage large drone fleets (i.e., drone light shows) or how air traffic controllers manage traffic, but with the critical difference that the operator is responsible for specific flights. Human factors engineering is crucial here—screens, alerts, and decision-support tools must enable one person to safely handle, say, five to ten aircraft at once in normal conditions and quickly intervene if one has a problem.

The Operations layer will need to develop standard operating procedures for all phases of flight that consider varying levels of human control. For example, if an autonomous eVTOL can't land at its intended vertiport due to a sudden storm, what is the procedure? Does it ask a supervisor for a new clearance, divert to an alternate, or hold? Accurate weather monitoring and forecasting models are essential to prevent such occurrences, especially in short-range/time trips at the early stages of the AAM implementation period. These procedures must be predefined and trained. There will also be regulatory requirements for rest and duty cycles—if one person is overseeing multiple flights, does that count the same as flying one plane for fatigue purposes? Certainly, authorities and the industry will need to work on these questions and find practical answers before the extended adoption of the AAM in our daily lives.

Fleet Management and Dispatch

Running an AAM service involves classic fleet management challenges, amplified by new tech. Operators must decide how many vehicles to station at each vertiport or region based on demand patterns (morning and evening peaks for commuters, midday lulls, etc.). Predictive analytics might be used to reposition aircraft in anticipation of ride requests (similar to how rideshare services reposition drivers). Because eVTOLs need charging, dispatch algorithms must incorporate battery state: an aircraft might only be offered a mission if it has sufficient charge plus reserve to complete it; otherwise, it must be routed to charging first.

Vertiport scheduling is another element, as there are only a certain number of vehicles that can land or take off at a vertiport per hour, given pad constraints, so the operational software needs to schedule slots (this ties into airspace management as well, ensuring flights to the same vertiport don't face any conflict). Some companies are developing integrated fleet operations centers, which will resemble a small version of airline operations centers—monitoring weather, airspace, vehicle status, vertiport status, and booking demands, then making decisions or feeding recommendations to pilots. For example, if high winds pick up at a particular vertiport, the operations center might pause departures there and communicate to passengers via app about delays, while sending additional vehicles to other vertiports as backups. All of this requires sophisticated software systems—essentially, the "back-end" IT infrastructure of AAM operations is as critical as the aircraft themselves. A failure in the dispatch software could cause significant disruption or even safety issues, so it will need redundancies and testing.

COMMUNITY

The community layer encompasses the human, social, and political factors that influence AAM's implementation. This includes public acceptance (are people willing to have and use AAM in their communities?), stakeholder engagement (educating and involving citizens, city officials, businesses), the development of regulations and policies (by governments responding to public interests), and considerations of how AAM will impact society at large (environmental effects, equity, noise, privacy, jobs). Many experts consider this layer the *make-or-break factor* for AAM: even if the technology works and business cases exist, AAM will not get off the ground (figuratively or literally) if society does not grant it a "social license" to operate. History provides lessons—for instance, earlier attempts at urban air mobility with helicopters faced public backlash over noise and a high-profile accident, which curtailed those services. AAM proponents are keenly aware of this and are actively working on community integration strategies (Rizzi et al., 2020).

Public Perception and Acceptance

Initial public surveys about AAM show a mix of excitement and concern. A notable study by EASA in 2021 surveyed urban residents in Europe and found that 83% of respondents had a positive initial attitude toward urban air mobility, and 71% said they would be willing to try an air taxi or drone service (EASA, 2021). This indicates a reservoir of curiosity and openness—likely fueled by the novelty and potential convenience. Support was especially strong for use cases seen as socially beneficial, such as emergency medical transport or humanitarian applications. However, alongside this optimism, the public expresses clear concerns that must be addressed:

Safety

Safety is the number one concern in virtually all surveys (EASA, 2021; Shaheen et al., 2018). People need to be confident that these new aircraft won't fall out of the sky or crash into their properties. Aviation is generally very safe, and AAM leaders often emphasize that they aim for airline-level safety. But until the public sees a track record, there may be skepticism. Any early accident could heavily damage trust (as happened with the helicopter era). Hence, community engagement often involves demonstrating safety testing and certification rigor.

Regulators play a role in reassurance, too: having clear, stringent certification standards and oversight helps signal to the public that AAM is not a tech wild west but under proper control (Cohen & Shaheen, 2021).

The community layer thus intersects with aircraft/operations here—achieving actual safety and transparently communicating it.

Noise

While eVTOLs can be much quieter than helicopters, they are not silent. The character of the noise (pitch, frequency, modulation) also matters for annoyance. Urban residents are wary of more noise pollution in already noisy environments. A drone-like buzzing might be perceived as more annoying than a distant airplane hum if it's persistent.

The EASA study found noise to be a major factor in acceptance. When playing sample AAM vehicle sounds for people, their willingness to accept frequent operations depended on the noise levels (EASA, 2021a). Many indicated that if noise is low enough to be similar to background traffic noise, it's acceptable, but if it's clearly noticeable or high-pitched, it could be an issue. There is also the phenomenon of multiple vehicles—one eVTOL might be fine, but if dozens are flying over an area daily, the cumulative effect could grate on people. Therefore, meeting noise targets is seen as crucial. This involves both technology (as discussed, quieter rotors) and operational measures (altitude restrictions, avoiding sensitive areas like schools or parks when possible, and perhaps limits on night flights). Some city authorities might impose noise quotas or require noise certification standards to be met (similar to how airports have noise rules for jets).

Privacy

The idea of vehicles flying over backyards or city streets raises privacy questions. Will these aircraft have cameras or sensors that could inadvertently (or deliberately) film people on rooftops or in private spaces? Even if the cameras are for navigation and not surveillance, the public might feel uneasy. Clear policies will be needed on data collection: e.g., assuring that any video sensor data is not stored or used to spy, and perhaps technical measures like downward-facing cameras that blur or avoid high resolution. Privacy also pertains to noise in a way—noise intrudes into the home environment.

Additionally, there's a psychological aspect, as some people simply dislike the thought of being "watched from above," a concern that grew with the advent of small drones. If AAM is seen as an extension of that, communities might demand certain airspace be off-limits. Some cities consider *drone no-fly zones* over residential areas for privacy, though generally, airspace is federally regulated in most countries (Kamal et al., 2023; Tian et al., 2019). Addressing privacy might involve education (explaining what tech is and isn't on the vehicles), regulations (like forbidding any non-navigational surveillance), and perhaps design (e.g., window tinting such that passengers can't easily zoom in on people on the ground).

Security and Safety (Ground Risk)

People also worry about the consequences of an accident or incident. What if an eVTOL makes an emergency landing on a street, or debris falls from the sky? The image of a vehicle crashing into a building is scary, even if the probability is extremely low. Community acceptance will depend on demonstrating that these scenarios are exceedingly unlikely and that, if they occur, there are mitigation measures (parachutes, designated emergency landing zones like open fields or parks, etc.). There's also concern about criminal misuse—could someone hijack a drone or eVTOL and crash it intentionally? Security agencies are actively considering these scenarios. As a result, initial operations might be quite constrained to mitigate risks, e.g., routes over water or highways rather than over crowds.

Community outreach can involve sharing emergency response plans with local authorities and even conducting open demonstrations of how an emergency would be handled, for example, showing how a forced landing can be done safely in a controlled way. Over time, as familiarity grows and incidents, hopefully, do not occur, public fear tends to abate. But at the outset, emphasizing safety margins—such as *"if anything goes wrong, the aircraft can automatically glide or parachute down to a safe landing"*—will be vital in messaging.

Visual and Environmental Impact

Some community concerns are broader: will the skyline now be dotted with moving vehicles where previously only birds flew? This visual change might bother some, especially in scenic or historic locales. It's somewhat subjective, but urban planners may consider limits on operations in certain treasured sightline areas, for example, to restrict eVTOLs from buzzing directly in front of the Eiffel Tower or Statue of Liberty regularly.

Environmentally, while eVTOLs are zero-emission at the point of use, people may ask about the broader lifecycle: e.g., battery production impacts, disposal, and the source of electricity (if it's coal power, then indirectly there are emissions). AAM companies often tout that their operations will be green, and many commit to using renewable energy for charging. Transparent environmental assessments will help—if studies show that replacing car trips with eVTOL flights reduces overall CO_2 or saves time, which can bolster public support among climate-conscious citizens (Goyal et al., 2021). On the other hand, if AAM is perceived as not actually environmentally beneficial or only serving a few, that could turn opinion.

Equity and Accessibility

A critical social question is who benefits from AAM. If it's only wealthy businesspeople avoiding traffic in expensive air taxis, public sentiment could sour, viewing AAM as a nuisance that privatizes the skies for the elite. To counter this, the industry and policymakers often stress a long-term vision of accessible air mobility—where costs come down and many people can use it, or where services like emergency response benefit everyone. Early on, companies might do outreach like offering some free flights to emergency workers or setting aside some capacity for public services, to show it's not just a luxury. Cities might impose requirements or incentives for operations in certain underserved areas or for pricing models that aren't exclusionary. This is similar to how some cities require ride-sharing to serve all neighborhoods, not just tourist areas.

There's also the aspect of job creation. AAM could create jobs, i.e., pilots, maintenance, vertiport staff, and companies can highlight local hiring or STEM initiatives in communities, which helps frame AAM as bringing opportunity (McNab, 2023). Ensuring diversity and inclusion in who operates and uses AAM will likely be part of community messaging—for instance, outreach to get a diverse pipeline of pilots and engineers, or making vertiports physically accessible for disabled passengers from the start.

Economic and Societal Impact

The Community layer also considers the broader impacts of AAM as follows:

Economic Development and Equity

Proponents argue that AAM can boost economic growth by creating new industries (vehicle manufacturing, new services) and by improving transportation efficiency (saving time, which has productivity gains). Cities might be attracted to being AAM early adopters to brand themselves as innovation hubs, potentially attracting tech investment. However, there's caution needed to ensure this doesn't widen inequality. If only high-income individuals use the service initially, the general public might question public support (like the use of public land or funds for vertiports). One strategy to address this is integrating AAM with existing public transit systems, perhaps offering combined tickets or including AAM routes in municipal transportation plans, thereby treating it as part of the mobility ecosystem rather than a standalone luxury. For instance, if an air taxi can be booked through the same platform as the city bus, and maybe city transit passes could offer a discounted air taxi leg in certain circumstances, it normalizes the service.

Environmental Sustainability

AAM's environmental impact will be closely watched by communities and regulators. On one hand, electric eVTOLs produce no tailpipe emissions and could be powered by renewable energy, helping cities meet climate goals (some cities have set targets for reducing vehicle miles traveled or converting trips to sustainable modes—AAM could be part of that if it pulls people out of single-passenger cars). On the other hand, if AAM primarily induces new travel (people flying who wouldn't have driven that trip, or flying further because it's faster), it could add to overall energy use unless well-managed. Environmental NGOs may scrutinize AAM projects, and their support or opposition can influence community perception. Early mover companies often publish sustainability reports claiming each eVTOL trip uses less energy or emissions than the car trip it replaces, etc., to build the case that AAM aligns with climate action. Independent studies (e.g., by universities) are also emerging to verify these claims. Ensuring that the life-cycle footprint of AAM (including battery production and disposal) is trending downward will be important for long-term acceptance, especially in communities that pride themselves on sustainability.

Quality of Life and Urban Form

In a positive sense, if AAM can reduce traffic congestion (by taking some trips off the road) or provide connectivity to underserved areas (e.g., connecting a remote suburb with jobs in the city quickly), communities may see an improvement in quality of life. This is part of the narrative AAM advocates push—that it's not just about tech, but about solving real mobility pain points. For example, a commuter spending 90 minutes in traffic might only spend 15 minutes in an air taxi; if that becomes affordable and routine, that person gains back hours of their day—a quality of life boost. However, skeptics worry it could also divert investment from ground infrastructure or lead to urban sprawl if people think they can live far away and still commute quickly by air (though high cost may limit that effect). Urban planners are discussing how AAM might fit into plans for more livable, walkable cities. It's a bit paradoxical—flying cars evoke sprawl, yet if used smartly, they could relieve roads and complement mass transit. Including community voices in those discussions is important to avoid technocratic decisions that ignore local context.

NOTES

1. Data used were best estimates, and the forecast is based on information available in the open public as of October 2025.
2. These are forecast values; real-world figures will vary with breakthroughs and raw material supply.
3. Combined Charging System (CCS) is a standardized charging interface used by many EVs worldwide.
4. Joby flew a hydrogen-electric demonstrator using an H_2 fuel-cell system developed with H2FLY and a liquid-hydrogen tank (Joby, 2022).
5. VOR/DME is a ground-based radio navigation aid that combines a VOR station with co-located Distance Measuring Equipment to give both bearing and distance to the same point.

REFERENCES

AeroTime. (2025, August 7). Joby confirms strong Q2 '25 cash position, certification progress, and Japan expansion. https://www.aerotime.aero/articles/joby-confirms-strong-q225-cash-position-certification-progress-japan-expansion

Al-Rubaye, S., Tsourdos, A., & Namuduri, K. (2023). Advanced air mobility operation and infrastructure for sustainable connected eVTOL vehicle. *Drones*, *7*(5), 319. https://doi.org/10.3390/drones7050319

Arafat, M. Y., & Pan, S. (2024). *Urban air mobility communications and networking: Recent advances, techniques, and challenges. Drones*, *8*(12), 702. https://doi.org/10.3390/drones8120702

Arup. (2022). *Air One: Designing the UK's first hub for eVTOL advanced air mobility.* Urban-Air Port. https://www.arup.com/projects/air-one/

BloombergNEF. (2023, November 26). *Lithium-ion battery pack prices hit record low of $139/kWh* [Press release]. https://about.bnef.com/insights/clean-energy/lithium-ion-battery-pack-prices-hit-record-low-of-139-kwh/

China Daily. (2025, April 22). Meituan secures first nationwide low altitude logistics operating license. https://www.chinadaily.com.cn/a/202504/22/WS6807752fa3104d9fd3820e14.html

Cohen, A. P., & Shaheen, S. A. (2021). Urban air mobility: Opportunities and obstacles. In *Encyclopedia of transportation* (pp. 702–709). Elsevier. https://doi.org/10.1016/B978-0-08-102671-7.10764-X

Cole, W., & Karmakar, A. (2023, June). *Cost projections for utility-scale battery storage: 2023 update* (NREL/TP-6A40-85332). National Renewable Energy Laboratory. https://www.nrel.gov/docs/fy23osti/85332.pdf

Conrad, C., Delezenne, Q., Mukherjee, A., Mhowwala, A., Ahmed, M., Zhao, J., … Tsourdos, A. (2023). Developing a digital twin for testing multi-agent systems in advanced air mobility: A case study of Cranfield University and Airport. In *2023 IEEE/AIAA 42nd digital avionics systems conference (DASC)* (pp. 1–10). https://dspace.lib.cranfield.ac.uk/server/api/core/bitstreams/89e4862b-c8cc-4f07-8079-47faf32d2cd5/content

Deniz, S., Wu, Y., Shi, Y., & Wang, Z. (2024). A reinforcement learning approach to vehicle coordination for structured advanced air mobility. *Green Energy and Intelligent Transportation*, *3*(2), 100157. https://doi.org/10.1016/j.geits.2024.100157

EHang Holdings Limited. (2023a, October 13). *EHang successfully obtains type certificate for EH216-S passenger-carrying UAV system issued by Civil Aviation Administration of China* [Press release].

EHang Holdings Limited. (2023b, December 21). *EH216-S passenger-carrying UAV system obtains standard airworthiness certificate from CAAC and the certified aircraft delivered to customer in Guangzhou* [Press release]. GlobeNewswire.

EHang Holdings Limited. (2024, April 7). *EHang secures production certificate from CAAC, clearing path for mass production of EH216-S pilotless eVTOL aircraft* [Press release].

Elsayed, M., & Mohamed, M. (2024). Robust digital-twin airspace discretization and trajectory optimization for autonomous unmanned aerial vehicles. *Scientific Reports*, *14*, 14035. https://doi.org/10.1038/s41598-024-62421-4

European Commission, Directorate-General for Mobility and Transport. (2023, January 26). *New EU rules on dedicated airspace for drones enter into force.*

European Union Aviation Safety Agency. (2021, May 19). *Study on the societal acceptance of Urban Air Mobility in Europe* [Report].

Federal Aviation Administration. (2023). *Advanced Air Mobility (AAM) implementation plan—Innovate28.* https://www.faa.gov/sites/faa.gov/files/AAM-I28-Implementation-Plan.pdf

Federal Aviation Administration. (2025, June 23). *New and emerging entrants on airports* https://www.faa.gov/airports/new_entrants

Flores, A., Ziakkas, D., & Delisle, J. (2023). The role of human-centered design in advanced air mobility implementation process. In *Proceedings of the AHFE international conference*. https://openaccess.cms-conferences.org/publications/book/978-1-958651-91-9/article/978-1-958651-91-9_36

Goyal, R., Reiche, C., Fernando, C., & Cohen, A. (2021). Advanced air mobility: Demand analysis and market potential of the airport shuttle and air taxi markets. *Sustainability, 13*(13), 7421. https://doi.org/10.3390/su13137421

Guo, J., Chen, L., Li, L., Na, X., Vlačić, L., & Wang, F.-Y. (2024). Advanced air mobility: An innovation for future diversified transportation and society. *IEEE Transactions on Intelligent Vehicles, 9*, 3106–3110. https://doi.org/10.1109/TIV.2024.3377464

Hays, K., Elangovan, A., Lopez, H., Sinkula, M., & Kumar, S. (2024). Ionblox's SiOx-based, high-energy, high-power Li-ion battery for performance electric vehicles and electric aviation. *ECS Meeting Abstracts*. https://doi.org/10.1149/ma2024-012504mtgabs

He, J., He, Q., Xu, Z., Jia, Y., Wang, J., Li, K., & Tan, W. (2024). Key technologies and upgrade strategies for eVTOL aircraft energy storage systems. *Journal of Energy Storage, 95*, 114402. https://doi.org/10.1016/j.est.2024.114402

International Energy Agency. (2023, April 26). *Global EV Outlook 2023: Catching up with climate ambitions*. https://www.iea.org/reports/global-ev-outlook-2023

Joby. (2022). Joby confirms revolutionary low noise footprint following NASA testing. https://www.jobyaviation.com/news/low-noise-footprint/

Johnson, W., & Silva, C. (2021). NASA concept vehicles and the engineering of advanced air mobility aircraft. *The Aeronautical Journal, 126*, 59–91. https://doi.org/10.1017/aer.2021.92

Kamal, A., Vidaurri, J., & Rubio-Medrano, C. (2023). No-fly-zone: Regulating drone fly-overs via government and user-controlled authorization zones. In *Proceedings of the 24th International Symposium on Theory, Algorithmic Foundations, and Protocol Design for Mobile Networks and Mobile Computing*. https://doi.org/10.1145/3565287.3617633

Kannan, K., Baculi, J., Lombaerts, T., Kawamura, E., Gorospe, G., Holforty, W., … Brown, N. (2023). A simulation architecture for air traffic over urban environments supporting autonomy research in advanced air mobility. In *AIAA SciTech 2023 forum*. https://arc.aiaa.org/doi/pdf/10.2514/6.2023-0895.c1

Kiesewetter, L., Shakib, K., Singh, P., Rahman, M., Khandelwal, B., Kumar, S., & Shah, K. (2023). A holistic review of the current state of research on aircraft design concepts and considerations for advanced air mobility applications. *Progress in Aerospace Sciences, 149*, 100949. https://doi.org/10.1016/j.paerosci.2023.100949

L.E.K. Consulting. (2021). Advanced air mobility—Cost economics and potential. https://www.lek.com/insights/ei/advanced-air-mobility-cost-economics-and-potential

Li, C., Zhao, Z., Chen, Y., Zhu, W., Qiu, J., Jiang, S., & Guo, R. (2024). Modeling the urban low-altitude traffic space based on the land administration domain model: Case studies in Shenzhen, China. *Land, 13*(12), 2062. https://doi.org/10.3390/land13122062

Lim, E., & Hwang, H. (2019). The selection of vertiport location for on-demand mobility and its application to Seoul metro area. *International Journal of Aeronautical and Space Sciences, 20*, 260–272. https://doi.org/10.1007/s42405-018-0117-0

Liu, J. (2023, June 12–15). *Progress and status of Battery500 Consortium Phase II* [Conference presentation slides]. U.S. DOE Vehicle Technologies Office Annual Merit Review. https://www1.eere.energy.gov/vehiclesandfuels/downloads/2023_AMR/bat317_liu_2023_o%20-%20jun%20liu.pdf

McNab, R. (2023). Advanced air mobility, economic impacts, and equity considerations. *Journal of Economic Analysis, 3*(2). https://doi.org/10.58567/jea03020009

Namuduri, K. (2023). Digital twin approach for integrated airspace management with applications to advanced air mobility services. *IEEE Open Journal of Vehicular Technology, 4*, 693–700. https://doi.org/10.1109/OJVT.2023.3312277

NASA. (2022a, October 10). NASA goes live with surrogate eVTOL flight tests in Texas. https://www.nasa.gov/centers-and-facilities/armstrong/nasa-goes-live-with-surrogate-evtol-flight-tests-in-texas/

NASA. (2022b). *Acoustic flight test of the Joby Aviation AAM aircraft (STRIVE S5).* NASA Technical Reports Server. https://ntrs.nasa.gov/api/citations/20220006729/downloads/Aeroacoustics2022_Pascioni_STRIVES5.pdf

NASA Aeronautics Research Mission Directorate. (2021). *Urban Air Mobility (UAM) Vision ConOps – UAM Maturity Level 4.* https://ntrs.nasa.gov/citations/20210010443

Rahman, B., Bridgelall, R., Habib, M., & Motuba, D. (2023). Integrating urban air mobility into a public transit system: A GIS-based approach to identify candidate locations for vertiports. *Vehicles, 5*(4), 97. https://doi.org/10.3390/vehicles5040097

Reuters. (2025, July 1). *Dubai aims to beat the traffic with 2026 Joby air taxi liftoff. Geo.tv.* (Syndicated Reuters item).

Rizzi, S. A., Huff, D. L., Boyd, D. D., Jr., Bent, P., Henderson, B. S., & Pascioni, K. A. (2020, October). *Urban air mobility noise: Current practice, gaps, and recommendations* (NASA/TP–2020-5007433). National Aeronautics and Space Administration.

Shaheen, S. A., Cohen, A. P., & Farrar, E. M. (2018). The potential societal barriers of urban air mobility (UAM) (NASA Contractor Report; TSRC Report No. G28C9TFR). UC Berkeley, Transportation Sustainability Research Center. https://escholarship.org/uc/item/7p69d2bg

Shenzhen Daily. (2024a, August 5). *City plans to add over 1,000 low-altitude flight routes by 2025.* Shenzhen Government Online.

Shenzhen Daily. (2024b, January 5). *Regulations to boost low-altitude economy.* Shenzhen Government Online.

Silva, C., Johnson, W., Antcliff, K. R., & Patterson, M. (2018). VTOL urban air mobility concept vehicles for technology development. In *AIAA Aviation 2018* (AIAA 2018-3847). https://doi.org/10.2514/6.2018-3847

Song, K., & Lee, H. (2025). Network topology-driven vertiport placement strategy: Integrating urban air mobility with the Seoul Metropolitan Railway System. *Applied Sciences, 15*(7), 3965. https://doi.org/10.3390/app15073965

Stubbins, G. (2025, November 19). *Joby eVTOL moves toward final stages of certification, builds global AAM framework. CompositesWorld.*

TechNode. (2025a, April 23). Meituan secures China's first nationwide drone delivery license, ramping up competition with Google Wing. https://technode.com/2025/04/23/meituan-secures-chinas-first-nationwide-drone-delivery-license-ramping-up-competition-with-google-wing/

TechNode. (2025b, July 31). Amap launches low-altitude navigation with 3D digital mapping in Shenzhen. https://technode.com/2025/07/31/amap-launches-low-altitude-navigation-with-3d-digital-mapping-in-shenzhen/

Thipphavong, D., et al. (2018). Urban air mobility airspace integration concepts and considerations. In *AIAA Aviation 2018* (AIAA 2018-3676). https://doi.org/10.2514/6.2018-3676

Tian, G., Xiao, C., & Liu, W. (2019). Dynamic no-fly zone for drones. In *2019 IEEE SmartWorld, Ubiquitous Intelligence & Computing, Advanced & Trusted Computing, Scalable Computing & Communications, Cloud & Big Data Computing, Internet of People and Smart City Innovation (SmartWorld/SCALCOM/UIC/ATC/CBDCom/IOP/SCI)* (pp. 761–766). https://doi.org/10.1109/SmartWorld-UIC-ATC-SCALCOM-IOP-SCI.2019.00163

Yang, X., Liu, T., Ge, S., Rountree, E., & Wang, C.-Y. (2021). Challenges and key requirements of batteries for electric vertical takeoff and landing aircraft. *Joule, 5*(7), 1644–1659. https://doi.org/10.1016/j.joule.2021.05.001

Zhang, Y. (2024, August 16). *City plans to add over 1,000 low-altitude flight routes by 2025.* EyeShenzhen.

Zhao, Y., & Feng, T. (2024). Strategic integration of vertiport planning in multimodal transportation for urban air mobility: A case study in Beijing, China. *Journal of Cleaner Production, 449,* 142988. https://doi.org/10.1016/j.jclepro.2024.142988

Ziakkas, D., St-Hilaire, M., & Pechlivanis, K. (2024). *The role of digital twins in the certification of advanced air mobility (AAM) systems.* In T. Ahram, W. Karwowski, D. Russo, & G. Di Bucchianico (Eds.), *Intelligent Human Systems Integration (IHSI 2024): Integrating People and Intelligent Systems* (pp. 476–485). AHFE Open Access. https://doi.org/10.54941/ahfe1004512

3 AAM—Present Applications

INTRODUCTION

On a spring night in Baltimore, a transplant team waited while a donor kidney traced a route few organs had taken before—lifted by a small unmanned aircraft across the city and handed, minutes later, to surgeons who restored a patient's chance at a normal life (Scalea et al., 2019; Talaie et al., 2021). That short flight captured the human stakes of advanced air mobility (AAM): when minutes matter, new aircraft, autonomy, and logistics can translate directly into lives saved. It also foreshadowed the questions that follow each successful sortie: Who is accountable when autonomy replaces a courier? What risks shift to communities under the flight path? And how do we scale such missions safely, equitably, and affordably?

AAM is already reshaping workaday tasks—urgent medical logistics, time-critical inspection, express freight, and piloted or remotely piloted passenger shuttles—by combining electric propulsion, distributed architectures, and increasing levels of automation. These technologies open new operational envelopes while compressing trip times and enabling novel service models (Kiesewetter et al., 2023). Yet a forward-looking assessment demands rigor: performance margins, energy density, reliability growth, and certification pathways remain determinative of which use-cases are genuinely "current" versus merely imminent.

At the same time, AAM's deployment is inseparable from governance and societal consent. Comparative policy research highlights the need for coherent regional frameworks that integrate technology, infrastructure, airspace, and adoption—rather than treating them in isolation (Raghunatha et al., 2023). Scoping reviews of public attitudes underscore recurring concerns—noise, visual intrusion, privacy, and data stewardship—that can erode acceptance if not confronted early with evidence, transparency, and mitigation (Wang et al., 2023). Legal scholarship continues to probe how long-standing air law principles should adapt to high-density, low-altitude operations and growing autonomy (Scott & Öykü, 2025).

Against that backdrop, this chapter focuses on what is flying now. We map the principal application families (medical logistics, emergency response, inspection and data services, cargo, and early passenger operations), then examine selected case studies of fielded programs. Emphasis is placed on practical deployments in the United States and China, with a survey of notable activity across Europe, Africa, and Australia.

DOI: 10.1201/9781003546306-4

TYPES OF AAM APPLICATIONS

AAM vehicles can support multiple use cases, each with distinct operational profiles and benefits. The most commonly discussed applications include on-demand passenger transport, cargo logistics, healthcare, and emergency services. By leveraging vertical takeoff and landing capabilities and high degrees of automation, these new air services promise to reduce travel times, bypass ground congestion, and improve access in remote areas (Bridgelall & Tolliver, 2024; Johnson & Silva, 2021). Recent studies also suggest substantial market demand if costs can be managed and operations integrated safely into existing transportation systems (Goyal et al., 2021).

Figure 3.1 illustrates the shift in the AAM application sectors from 2015 to 2025, and a forecast for 2035. It is clear that the main volume has been achieved by thousands of drones annually by delivering goods and supporting the healthcare services, with the passenger transport being the least developed yet, and with many struggles that need to be overcome.

Below, we outline key AAM application domains and their current status.

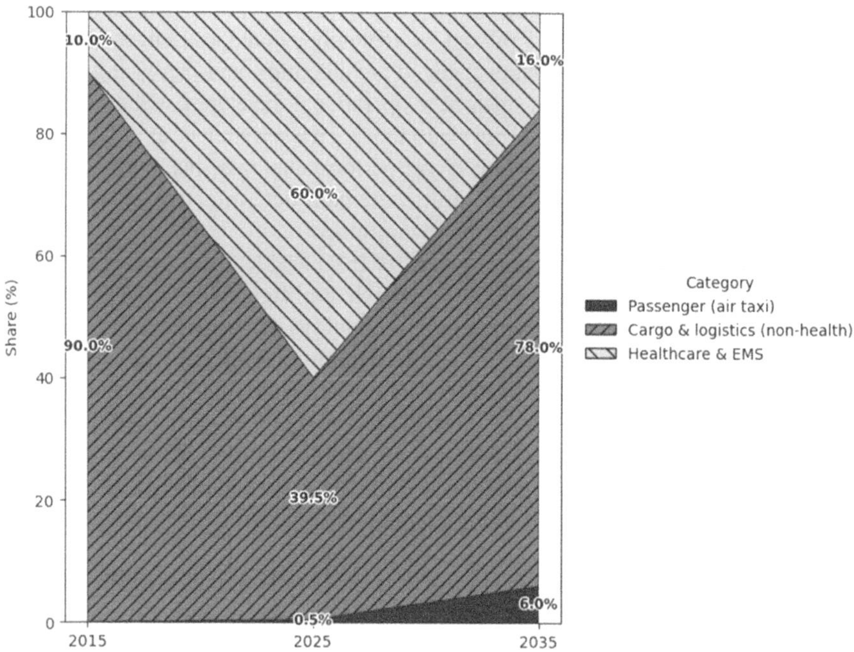

FIGURE 3.1 AAM application sectors by volume. (Credit: Author, Data source: Joby Aviation, 2024b, 2025; Skyports Infrastructure, 2024; Advanced Technology Research Council, 2025; Reuters, 2025; Archer Aviation, Inc., 2024, 2025; The Verge, 2025; Zipline, 2024; McKinsey & Company, 2023, 2024; PwC, 2024; Wing, 2025; DroneDJ, 2024; DroneXL, 2025; Deutsche Post DHL Group, 2014; Swiss Post, 2017; EHang, 2023, 2025.)

Passenger Transport and Air Taxis

Passenger transport is one of the most prominent AAM applications. Electric vertical takeoff and landing (eVTOL) air taxis offer the potential to reduce travel times in congested urban environments and to link remote or suburban areas with major hubs (Bridgelall & Tolliver, 2024; Johnson & Silva, 2021). The iconic vision is a network of small eVTOL *air taxis* ferrying passengers on demand, much like app-based ride-sharing, but in the sky (see Figure 3.2). Initial operations are expected to be short (e.g., 10–30 minute hops) and possibly along predetermined corridors between vertiports in high-demand locations. For example, Archer Aviation, in partnership with United Airlines, announced plans in 2023 for a Chicago Airport Shuttle route—using eVTOLs to connect downtown Chicago (Vertiport Chicago) to O'Hare International Airport in ~10 minutes, a trip that often takes an hour by car (Archer Aviation, Inc., 2023). This service, slated for 2025, would operate high-frequency during peak travel times and demonstrate the value of bypassing ground traffic for airport transfers.

Early use cases like airport shuttles are seen as potentially viable segments for AAM, since travelers may pay a premium to save time when catching. In general, recent demand analysis underlines significant market potential for urban air taxis and airport shuttles, estimating that these services could serve around 82,000 passengers daily in the United States, if widely implemented, translating to an annual market of roughly $2.5 billion (Goyal et al., 2021). By reducing ground congestion and capitalizing on simplified vehicle operations, eVTOL concepts are quickly transitioning from design to operational trials (Wing et al., 2020). Several major eVTOL developers in the United States (e.g., Joby Aviation, Archer, Beta Technologies) have conducted extensive prototype test flights and are working with regulators toward certification and commercial launch. These companies have also pursued Part 135

FIGURE 3.2 Passenger AAM trial flight. (*Source:* EHang.)

air carrier certificates to lay the groundwork for operating air taxi services (Archer Aviation, Inc., 2024; Federal Aviation Administration, 2022).

In parallel, public perception and community integration (e.g., noise management, vertiport placement) are being addressed through demonstration flights and outreach.

Overall, passenger air taxis represent the flagship AAM use case, and although routine services were not yet operational as of 2025, multiple pilot programs and route announcements signal that this application is on the tipping point of becoming a reality.

Cargo and Logistics

Another promising AAM application is uncrewed or remotely piloted cargo delivery. This ranges from small-package drone delivery, for consumer goods or medical supplies, to larger payload transport using bigger eVTOL or hybrid aircraft (see Figure 3.3). By leveraging lower-altitude airspace, autonomous cargo drones can significantly shorten delivery times for time-sensitive goods while reducing dependence on ground transportation infrastructure (Moradi et al., 2024). Companies like Amazon (through its Prime Air initiative) and Alphabet's Wing have been developing autonomous delivery drones to transport packages directly to customers' yards or designated drop-off points.

In parallel, logistics firms (UPS, DHL) and startups are trialing cargo-specific eVTOLs designed without passengers, maximizing space for parcels. Importantly, these operations hinge on advanced flight autonomy, collision-avoidance systems, and robust unmanned traffic management (UTM) to ensure safe integration with other airspace users (Chin et al., 2023; McCorkendale et al., 2024). Some current deployments highlight the potential: Wing's Drone Delivery Service, for instance, has operated in suburban communities in Australia and the United States, at times delivering over 1,000 packages per day (up to one delivery every 25 seconds during peak demand) in pilot markets (Stonor, 2023). Likewise, China's Meituan has integrated drones into its food delivery platform, reportedly completing about 170,000 drone orders across 18 urban communities since 2021 (Cornell et al., 2023). These trials indicate that drone logistics can augment last-mile delivery networks, especially for lightweight items.

Cargo AAM operations can potentially run 24/7—unlike passengers, "cargo doesn't sleep"—which improves vehicle utilization, though noise regulations may restrict night flights in populated areas. Key challenges for cargo drones include package security (ensuring items are not lost or damaged, and drop-off points are safe) and public acceptance of low-flying vehicles. However, the benefits of bypassing road congestion for urgent deliveries (e.g., medicines, organs for transplant) are significant. Notably, during the COVID-19 pandemic, medical drone deliveries were deployed in some regions to transport test samples and vaccines rapidly, underscoring the life-saving value of this AAM application (Moradi et al., 2024). As regulations evolve (e.g., enabling beyond-visual-line-of-sight operations) and autonomous technology matures, cargo and logistics applications of AAM are expected to scale up rapidly.

FIGURE 3.3 Google's drone delivery company, Wing, partnered with DoorDash flies through the sky to deliver food in Charlotte, NC, United States. (*Source:* Shutterstock.)

Healthcare and Emergency Services

The healthcare sector stands to benefit greatly from AAM, through applications such as medical deliveries, aerial emergency response, and air ambulances (Bridgelall & Tolliver, 2024; Goyal & Cohen, 2022). Drones are already being used to carry blood products, vaccines, and medications to clinics in areas with poor road access, dramatically reducing delivery times and improving healthcare outcomes. For example, in parts of sub-Saharan Africa, drone networks operated by companies like Zipline have become an essential part of the medical supply chain—in Rwanda, drones now deliver 75% of blood supplies outside the capital city, reaching remote hospitals in minutes rather than hours (Leedom, 2024). This has corresponded with measurable health benefits; one study reports a 51% reduction in postpartum hemorrhage deaths in Rwanda due to the rapid blood deliveries by drones (Leedom, 2024).

Emergency response is another critical use case (see Figure 3.4). Medical emergency drones can be dispatched to bring defibrillators or EpiPens to the scene of an incident before paramedics arrive, or even to transport a doctor or an emergency medical technician via eVTOL to a hard-to-reach location. Several pilot projects are exploring eVTOL air ambulances that could evacuate patients from accident sites or rural areas straight to hospitals, bypassing ground traffic delays (Wheeler et al., 2024). For instance, in Europe, the non-profit ADAC Luftrettung (the largest helicopter ambulance operator in Germany) has partnered with Volocopter to study using eVTOLs for emergency medical services; test flights are planned to begin in 2025 to evaluate response time improvements (Nevan, 2024, as cited in Designboom, 2024).

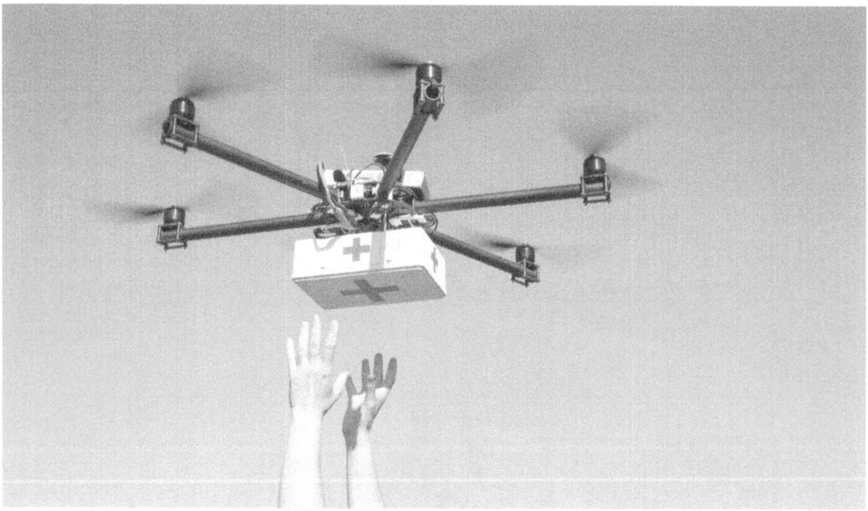

FIGURE 3.4 First responder drone carrying medicine package, person receiving first aid help, Munich, Germany, 14 April 2024. (*Source:* Shutterstock.)

During natural disasters or search and rescue (SAR) missions, unmanned aircraft can also play a vital role by surveying damage, locating survivors, or delivering relief supplies to areas cut off by debris. Small drones were deployed after hurricanes and earthquakes in recent years to assist first responders (Wheeler et al., 2024). Likewise, fire departments have experimented with drones for situational awareness in large fires, and companies are prototyping larger drones that could drop water or retardant on wildfires and high-rise building fires (see Figure 3.5). A prominent example is Chinese manufacturer EHang's AAV for firefighting, which has been demonstrated to carry firefighting materials to high floors where conventional fire trucks cannot reach.

FIGURE 3.5 Illustrative example of an autonomous aerial vehicle spraying foam to extinguish fire on a high-rise building. (*Source:* Shutterstock.)

Overall, AAM in healthcare and emergency services offers some of the clearest public benefits, which is reflected in public opinion surveys in Europe have found strong support for AAM when used for medical or emergency purposes (European Union Aviation Safety Agency, 2021). While these operations do not always follow a commercial business model (many are government or NGO-led), they provide a crucial test bed for technology and help build community acceptance of advanced air vehicles by saving lives.

GLOBAL AAM DEVELOPMENTS AND CASE STUDIES

To understand the current trajectory of AAM, it is instructive to examine how different regions of the world are implementing these new air mobility solutions. This section highlights case studies and initiatives from key regions—China, the United States, Europe, Africa, and Australia—illustrating diverse strategies and milestones. We focus on practical deployments and regulatory progress as of 2024–2025, noting that some regions have achieved limited commercial operations (especially in cargo and drone delivery) while others are still in the trial and demonstration phase.

Importantly, the level of AAM activity varies. China and the United States host the largest concentration of AAM companies and pilot programs, whereas regions like Africa and Australia have carved out specific niches, notably drone delivery in Africa and Australia's suburbs. Europe has strong industry players and government support, but faces a slower regulatory timeline.

United States

In the United States, AAM activity is driven largely by private sector innovation, with a strong emphasis on piloted eVTOL development and drone delivery pilots. The United States hosts several of the world's best-funded AAM startups and initiatives, though regulatory approvals are moving more cautiously than in China. The Federal Aviation Administration (FAA) has been working on integrating AAM through programs like the AAM National Campaign (led by NASA) and the BEYOND program for drone integration, focusing initially on safety frameworks. As of 2025, no passenger-carrying eVTOL had been certified in the United States, but progress is steady toward that goal. Two leading eVTOL companies—Joby Aviation and Archer Aviation—have achieved significant milestones on the path to commercial service. Joby, which is developing a five-seat piloted eVTOL, became the first company to complete stage 1 of FAA certification testing in 2023 and also received a Part 135 air carrier certificate (allowing it to operate an airline service) (Joby Aviation, 2024a; Polek, 2022). Archer similarly obtained its Part 135 certificate in 2024 and has been conducting intensive flight tests of its *Midnight* eVTOL prototype. Both companies target a 2025 or 2026 launch of air taxi operations in select cities, pending final type certification of their aircraft. United Airlines has placed orders and made investments in both Archer and a separate eVTOL developer (Eve Air Mobility), signaling airlines' interest in feeder and airport shuttle routes. As noted earlier, Chicago and New York are among the inaugural cities announced for Archer's and United's UAM network, with the Chicago airport shuttle route planning to start by 2025

FIGURE 3.6 Wisk Aero Gen6 aircraft. (*Source:* Wisk Aero.)

(Archer Aviation, Inc., 2023). United and Archer envision a hub-and-spoke model: initial "trunk" routes like O'Hare-to-downtown will build demand and familiarity, followed by "branch" routes connecting to suburbs.

Another major U.S. player, Wisk Aero (backed by Boeing and Kitty Hawk), is developing a fully autonomous air taxi (the Cora), as illustrated in Figure 3.6. Wisk's design is for a self-flying, four-seat eVTOL with no pilot onboard, a bold approach that will require new regulatory standards. In 2022, Wisk became the first to formally submit an autonomous eVTOL for FAA certification, but entry into service will likely lag behind piloted competitors (Brinkmann, 2022). Meanwhile, Beta Technologies is pursuing an electric aircraft that blurs the line between eVTOL and conventional flight—its ALIA vehicle can take off vertically or on a runway and is aimed at cargo and regional passenger markets (Bristow Group Inc., 2025). Beta has attracted interest from logistics companies (UPS has an order for cargo versions) and even the U.S. Air Force (which is evaluating it for transport missions). A diverse ecosystem of smaller startups (e.g., Lift Aircraft, which is testing single-seat recreational eVTOLs, and Overair in California) further highlights the United States as a hotbed of AAM R&D.

In the drone delivery arena, the United States has seen a blossoming of pilot programs, often in partnership with major retailers. Since 2021, retail giant Walmart has teamed up with drone operators (including DroneUp, Flytrex, Zipline, and Wing) to offer aerial delivery from dozens of stores in states like Arkansas, North Carolina, Texas, and Arizona. By late 2022, Walmart had established 36 drone-delivery hubs across seven states, collectively completing thousands of deliveries of items like groceries and home supplies (Cornell et al., 2023). These operations remain limited to a short radius (typically 1–2 miles from the store) and are conducted under FAA Part 107 drone rules with waivers to fly beyond visual line of sight (BVLOS). Initial customer response has been positive for speed and novelty, though scaling beyond test locales will depend on regulatory clearance for routine BVLOS flights. Amazon's Prime Air program, after years of development, launched drone deliveries in two small communities (in California and Texas) in late 2022. Amazon's hexagonal *MK27-2* drones have been delivering select packages (under 5 lb) from company warehouses directly to customers' backyards, with the drones descending to a safe height and dropping the package by tether. By mid-2023, Amazon had begun rolling out a newer drone model (MK30) and announced intentions to expand service to additional U.S. cities (Amazon Staff, 2023). However, volumes remain modest as Amazon works through technical and logistical challenges (such as noise, community acceptance, and a limited set of deliverable products).

The most prolific U.S. drone delivery operator to date is Alphabet's Wing, which started delivering in test markets like Christiansburg, Virginia, in 2019. Wing has since expanded to suburban Dallas–Fort Worth, Texas, and by 2023, it was completing tens of thousands of deliveries in those areas. In one reported week, Wing handled over 1,000 orders per day in its Dallas suburbs service, indicating robust demand when available (Stonor, 2023). Wing's strategy has involved partnering with local retailers and food chains (e.g., Walgreens, CVS, and in Australia, supermarket chain Coles) and positioning "nests" of drones that can be summoned via a smartphone app. The FAA has granted Wing certain waivers that allow it to operate with fewer human observers, leaning on its automated flight navigation and detect-and-avoid systems. Notably, Wing joined with DoorDash in 2022 to pilot drone delivery within the DoorDash app in Queensland, Australia (a model that could be extended to the United States for food delivery) (Cornell et al., 2023). Regulatory progress is accelerating: in 2023, the FAA approved several high-profile BVLOS drone operations without the need for visual observers, including for Zipline (which is expanding medical deliveries in North Carolina) and UPS Flight Forward (Cornell et al., 2023). These approvals suggest the FAA is gradually becoming more comfortable with autonomous flight in controlled scenarios, a positive sign for AAM growth.

At the municipal and state level, U.S. cities are also preparing for AAM. Los Angeles, for example, launched an *Urban Air Mobility Partnership* to study route networks and community impact, aiming to be one of the first adopters of air taxis around the 2028 Olympics (Daleo, 2024). Orlando, Florida, entered an agreement with German air taxi firm Lilium to establish a vertiport and routes in central Florida, with plans for a regional hub by the late 2020s (Lake Nona Editorial Team, 2020; Orlando Economic Partnership, 2020). New York City has seen interest from Blade (a helicopter shuttle company) to eventually incorporate eVTOLs on its routes to airports and the Hamptons, in partnership with manufacturers like Beta and Eve. In 2022, NASA and FAA jointly released an AAM Implementation Plan outlining a staged introduction of urban air mobility by first using existing heliports and underused regional airports for early eVTOL operations (Federal Aviation Administration, 2022). This approach would minimize new infrastructure needs and ease regulatory acceptance by leveraging proven flight procedures (since initially these eVTOLs will fly much like helicopters, with licensed pilots and under instrument flight rules as required).

Public sentiment in the United States is cautiously optimistic but will depend on demonstrated safety and noise management. The community pushback that plagued some early drone trials—for instance, Wing's tests in the suburbs of Canberra, Australia, saw complaints about buzzing noise disturbing residents and pets—serves as a cautionary tale for U.S. deployments. The FAA has identified noise as a key factor and is working with industry to ensure eVTOL noise profiles are lower and less annoying than traditional helicopters (many eVTOL designs aim to be 10–15 dB quieter). So far, the most tangible AAM benefits in the United States have come via drones in healthcare logistics: partnerships like Zipline with hospitals in North Carolina and Arkansas have enabled on-demand delivery of blood, lab samples, and medical products, often cutting delivery times by 75% compared to couriers (McNab, 2023). These services, though small in scale, demonstrate clear value and have encountered minimal opposition, thus often serving as the vanguard of AAM introduction to communities.

The United States' AAM landscape in 2025 is one of active experimentation and gearing up for launch. Significant venture capital, over 50% of global AAM investment (*Zag Daily*, 2025), has flowed into American AAM ventures, and the country's large aviation market offers fertile ground for a variety of use cases from urban air taxis to rural drone deliveries. The next few years are expected to bring the first commercial eVTOL flights (with pilots) on U.S. soil, likely in a handful of major cities, as well as the scaling up of drone delivery networks as regulations permit. The FAA is laying important groundwork via advisory committees and small-step approvals. If the industry can navigate certification hurdles and public acceptance, the United States could see AAM become a regular feature of its transportation system by the late 2020s—offering a new mode of mobility in both metropolitan and remote communities.

China

China has rapidly emerged as a global leader in AAM, propelled by proactive government support, an agile regulatory approach, and robust industry investment. Under the national strategy of fostering a *low-altitude economy,* Chinese authorities have embraced AAM as an economic growth area and are fast-tracking the development of autonomous eVTOL services (*Urban Air Mobility News*, 2025). A landmark achievement came in October 2023 when the Civil Aviation Administration of China (CAAC) granted the world's first type certificate for a passenger-grade eVTOL—to EHang's model EH216-S, an autonomous two-seat aerial vehicle (EHang, 2023; Sampson, 2023). This was followed by a production certificate in early 2024, and in March 2025, CAAC issued the first-ever air operator certificates for pilotless aerial vehicles to EHang's operating subsidiaries (Kang, 2025). With these approvals, EHang began the world's first commercial eVTOL passenger operations: Chinese consumers can now buy tickets for short sightseeing flights in designated cities, initially Guangzhou and Hefei, marking the official start of paid, autonomous urban air mobility services (Kang, 2025). The early routes are focused on tourism and aerial views, often flying a circular route from a vertiport and back (Point A to A) to manage risk. These flights are monitored by ground control centers, but notably have no pilot on board—a stark contrast to Western regulators' insistence on piloted aircraft in initial AAM deployments (Ison, 2025). By achieving a full set of certifications (type design, production, and operation) for an unmanned eVTOL, China has leapfrogged ahead in enabling true autonomous air taxis. EHang's EH216-S had logged over 40,000 test flights by the time of certification, and the company reports strong demand from provincial governments and tourism zones for building *drone taxi* routes (EHang, 2025a, b).

Complementing these vehicle-centric advances, Chinese cities are investing heavily in AAM infrastructure and pilot programs. The city of Shenzhen offers a prominent case study as it aims to become a model *UAM city*. By 2023, Shenzhen hosted more than 1,700 drone and AAM-related enterprises, with a *low-altitude economy* output nearing ¥96 billion (~$13 billion) (*Beijing Review*, 2024). The municipal government has committed ¥12 billion (~$1.7 billion) through 2026 to build a comprehensive network of vertiports, drone launch pads, and communication systems for urban air mobility (Deng, 2024). Plans call for constructing over 1,200 takeoff and landing platforms across the city for eVTOLs and delivery drones (*Urban Air Mobility News*, 2024). As of late 2023, Shenzhen had already built 249 such facilities (atop buildings and in open spaces), aiming for 400+ by 2024 and hundreds more by 2025 (*Urban Air Mobility News*, 2024). Alongside physical infrastructure, Shenzhen is deploying an advanced digital UTM architecture. A new *Smart Integrated Lower Airspace System*—essentially an automated air traffic management "brain" for altitudes below 500 meters—is being rolled out to coordinate the expected surge in drone and eVTOL flights (Kwok, 2024). This builds on an existing UTM system that already handles over 2,000 drone flights per day in Shenzhen (Butterworth-Hayes, 2019). Notably, Shenzhen was the first Chinese city to pass specific local legislation for the *low-altitude economy*, effective early 2024, providing a legal framework for drone deliveries, passenger drones, aerial tours, and other AAM activities within the city (Shenzhen Daily, 2024). By establishing no-fly zones, flight approval processes, and liability rules, the city's regulations balance innovation with safety and privacy concerns.

Several pioneering use cases have been demonstrated in China. In February 2024, Shanghai-based startup AutoFlight conducted the world's first inter-city eVTOL air taxi flight, flying its five-seat Prosperity I craft from Shenzhen to the neighboring city of Zhuhai, across the Pearl River Delta. The unmanned, 50 km journey was completed in just 20 minutes—a trip that would take about 3 hours by car—showcasing the potential of regional AAM links (AutoFlight, 2024a). This demonstration route is part of a planned network of 100+ eVTOL routes in the Greater Bay Area, where local governments envision thousands of vertiports eventually connecting cities in Guangdong, Hong Kong, and Macau (AutoFlight, 2024a).

These early operations help familiarize the public with eVTOL rides and gather data on noise and reliability. Public acceptance appears cautiously optimistic, especially when services are framed as novel attractions or time-saving transit for well-understood needs, like airport shuttles. Surveys by Chinese researchers indicate that urban residents are willing to try air taxis for urgent trips, provided safety is assured (SAFE (Coalition for Reimagined Mobility), 2025).

China is also leveraging AAM for cargo and delivery applications at scale. The tech giant Meituan has seamlessly integrated drones into daily life in cities like Shenzhen, delivering food, groceries, and parcels through its on-demand delivery app. Meituan's drones operate along fixed routes from rooftop *launch sites* to pickup kiosks in residential areas. Impressively, as of 2024, Meituan had completed over 360,000 drone deliveries across several major cities (*Beijing Review*, 2024), demonstrating year-round operation in various weather conditions. Each drone flight carries small orders (meals, beverages, medicine) up to ~2.5 kg, typically covering a few kilometers in 5–10 minutes—significantly faster than ground delivery in dense traffic. The company has opened 40+ drone routes in Shenzhen, Beijing, Shanghai, and elsewhere, and its drones are designed to handle moderate rain, wind, and night operations (*Beijing Review*, 2024). Shenzhen alone recorded over 600,000 drone delivery flights in 2023, reflecting how routine this service is becoming in some Chinese cities (*Beijing Review*, 2024). To manage airspace, Meituan employs a proprietary UTM system that coordinates its fleet and communicates with city regulators. The drones largely fly along pre-approved low-altitude corridors, and authorities have designated certain "drone lanes" in the sky to segregate them from manned aircraft. In early 2024, Shenzhen's local law on unmanned aviation formally recognized such drone delivery operations, giving them clearer legal status (China Briefing, 2024). Other companies, like JD.com and Alibaba, have also trialed delivery drones, especially in less urban areas, and logistics firms are eyeing larger cargo eVTOLs for inter-warehouse transport. For instance, AutoFlight is developing a cargo drone (the *CarryAll*), which in 2023 became the first cargo eVTOL above one ton to receive a type certificate from CAAC (AutoFlight, 2024a). This cargo model is intended to carry heavier loads between logistics hubs, complementing the smaller last-mile drones.

China's distinct approach to AAM is characterized by top-down coordination and rapid iteration. The central government's 14th Five-Year Plan (2021–2025) explicitly calls for accelerating civil unmanned aviation and AAM infrastructure (*Urban Air Mobility News*, 2025). More than 30 pilot zones have been established nationwide to test AAM technology under varying conditions, from the mountainous provinces (exploring medical deliveries to remote villages) to metropolitan areas (testing passenger drones). Substantial public funding—an estimated $16 billion by local authorities—has been allocated for vertiport construction, air traffic control upgrades, and industrial parks devoted to eVTOL manufacturing (Cirium Ascend Consultancy, 2025). Chinese eVTOL companies benefit from strong domestic supply chains (for batteries, motors, composites) and ample venture capital. In fact, as of late 2024, Chinese companies like EHang, AutoFlight, and Geely's Aerofugia accounted for a dominant share of global eVTOL pre-orders, with Asia-Pacific overtaking North America in total eVTOL orders (*Zag Daily*, 2025). These firms are pursuing both domestic deployment and international market entry—for example, EHang has partnerships to launch aerial tourism partnerships (example: Bangkok/Phuket via Reignwood) (EHang, 2025c), and AutoFlight delivered its first Prosperity eVTOL to a customer in Japan in 2024 (AutoFlight, 2024b; Ros, 2024).

All in all, China's strides in AAM provide a valuable blueprint. Its combination of agile regulation, state-industry collaboration, and public investment has accelerated AAM implementation in ways few predicted a few years ago. As one analyst noted, "China exemplifies a high-speed, government-driven strategy that is bringing autonomous urban air transport into daily life" (SAFE (Coalition for Reimagined

Mobility), 2025). If the current trajectory continues, the late 2020s could see dozens of Chinese cities with operational AAM networks, offering a real-world glimpse of the aerial mobility future.

Europe

Europe is an active player in AAM, with strong backing from both industry and government, although its approach prioritizes safety and sustainability, resulting in a generally more conservative timeline. The European Union, through the European Union Aviation Safety Agency (EASA) and various funding programs, has issued comprehensive proposed regulations for eVTOL certification and urban air mobility concept of operations. Several European companies are among the pioneers of AAM technology: Germany's Volocopter and Lilium, and the United Kingdom's Vertical Aerospace are notable examples. As of 2025, none of these had started commercial service, but Europe has seen a number of high-profile demonstrations and pilot projects signaling progress.

Volocopter, in particular, has positioned itself to potentially launch one of the first eVTOL air taxi services in Europe. Its two-seat VoloCity air taxi received special authorization from French authorities to conduct test flights in Paris airspace during the run-up to the 2024 Olympic Games. In August 2024, Volocopter completed a crewed test flight at the Paris Saint-Cyr aerodrome vertiport, marking the first-ever eVTOL flight at a dedicated commercial vertiport in Europe (Designboom, 2024). This flight, with a pilot onboard, was carried out under a permit from the French DGAC (civil aviation authority) even though the vehicle had not yet obtained full type certification from EASA (Designboom, 2024). It was a significant step in operational validation: the Volocopter team practiced vertiport operations, ground handling, battery charging, and integration with air traffic control. While initial hopes to offer passenger flights for the public or dignitaries during the 2024 Olympics proved too ambitious—the VoloCity was not certified in time to shuttle Olympic spectators—the company still managed to fly demonstration sorties during the Olympic period, including a publicized flight over the gardens of the Palace of Versailles in late summer 2024 (Daleo, 2024). These flights were important to build confidence and gather data. Volocopter and its partner Groupe ADP (which operates Paris airports) constructed five vertiport sites around the region (including one on a barge on the Seine River) to form a trial network. Although full passenger service is pending certification, Paris continues to prepare for AAM: local officials aim to have routes linking Charles de Gaulle Airport, Le Bourget Airport, downtown Paris, and Versailles once safety approval is achieved (Daleo, 2024). Volocopter's CEO stated that after achieving EASA type certification—anticipated by 2025—the plan is to start with tourist flights and emergency medical flights (in partnership with the Paris hospitals and the ADAC Luftrettung in Germany) to gradually integrate AAM into public services (Designboom, 2024). Indeed, ADAC in Germany has agreed to purchase a small fleet of VoloCity eVTOLs to test air rescue operations; these will be used in a pilot program for airborne first response and inter-hospital transport, expected to begin trial runs in 2025–2026 (Daleo, 2024).

Meanwhile, Lilium, another German company developing a 6–7 seat eVTOL, faced financial and technical challenges that delayed its timeline. Lilium's unique ducted-fan aircraft aimed for longer-range regional flights, but as of 2024, the company had not yet achieved crewed flight tests of its production design and was seeking new investment (*Zag Daily*, 2025). This led to some consolidation in Europe's AAM scene, with Volocopter taking a clearer lead. The European Commission and national governments have been very supportive of AAM research: the EU framework programmes (Horizon 2020 via SESAR JU) has funded projects like CORUS and AMU-LED to simulate and test UAM operations in urban environments (EUROCONTROL, 2021; EUROCONTROL, n.d; European Commission, 2020). Furthermore, several European cities are part of the UAM Initiative of the EU's Smart Cities Marketplace, exchanging best practices on vertiport planning and social acceptance. As of 2023, at least 20 cities in Europe had active AAM working groups or pilot initiatives (Unmanned Airspace, 2023). For example, Hamburg, Germany, has tested drone deliveries of medical samples between hospitals (Hamburg Aviation, 2025), and London, United Kingdom, formed an AAM consortium to explore vehicle integration and landing site requirements, with trials planned in the Midlands and Scotland using small cargo eVTOLs (Pritchard, 2024). Additionally, the European Space Agency has been involved in developing UTM services that rely on satellite communication and navigation to support drone and eVTOL flights in remote areas, which could particularly benefit countries with dispersed geographies like Norway, Sweden, and Finland.

Regulatory coordination in Europe is a relative strength: EASA is creating common standards that will apply across EU member states, meaning an eVTOL certified by EASA could operate in any member country (plus likely the United Kingdom, which often mirrors EASA rules, and other aligned countries). EASA published special condition guidelines for VTOL aircraft in 2019 and a detailed Prototype Technical Design Specifications in 2020, addressing airworthiness requirements for eVTOLs. It has also been working on pilot licensing frameworks (including a new category for eVTOL pilots) and operational rules for UAM services. A crucial hurdle is type certification—as of mid-2025, no eVTOL had yet been type-certified by EASA, though Volocopter's VoloCity and Vertical Aerospace's VX4 were in the advanced stages of testing. The expectation is that industry and regulatory developments point toward mid-decade entry-into-service timelines, contingent on certification progress (Alcock, 2025; European Union Aviation Safety Agency, 2025; Warwick, 2025). The delay in meeting the optimistic Paris 2024 goal illustrates the high bar for certification: even with political will, the technical proof and safety case for these new aircraft must be watertight. European regulators are particularly attuned to safety and noise—the latter is an important public concern in Europe's often dense cities. However, the collaborative approach means that once approved, AAM services might ramp up quickly across multiple countries.

In terms of drone delivery, Europe has pockets of progress but overall lags behind the United States and China in practical deployment. Countries like Ireland and Iceland were early movers (e.g., Irish startup Manna has been delivering food and small goods in suburban Dublin since 2020, completing over 100,000 deliveries by 2022) (Cornell et al., 2023). The United Kingdom had some trials (Tesco grocery deliveries by drone in 2021), and Finland partnered with Wing for a pilot service in Helsinki back in 2019. Yet, European drone delivery activity remains relatively small-scale. One reason is the complex airspace and jurisdictional environment—coordinating drone rules across EU states and addressing privacy laws (especially stringent in Europe) is challenging. That said, interest is growing: in July 2022, the U.K. government announced investment supporting "drone superhighways," including the planned 165-mile Project Skyway corridor linking Reading, Oxford, Milton Keynes, Cambridge (and onward to Coventry and Rugby) (Department for Business, Energy & Industrial Strategy et al., 2022; UK Research and Innovation, 2025). In Switzerland, the drone company Matternet ran a multi-year program transporting lab samples between Zurich hospitals, and Switzerland's civil aviation authority has been a pioneer in authorizing routine BVLOS drone ops. African drone successes have not gone unnoticed in Europe—for example, Germany's development agency funded trials to use Wingcopter drones for medical deliveries in Malawi and is exploring similar models that could be applied in European rural areas (Leedom, 2024).

Europe's AAM ecosystem is characterized by methodical progress and strong public-sector involvement. While the United States and China race ahead in terms of near-term commercial operations, Europe may be setting the stage for a more sustainable and socially accepted rollout, even if a bit later. The commitment to use cases beyond just passenger taxis—such as regional connectivity (e.g., connecting smaller cities by eVTOL shuttles) and emergency services—could prove wise in demonstrating value. By 2030, the Commission envisages large-scale Innovative Air Mobility (including UAM) in Europe—targeting regular services across at least 45 EU cities/regions—supported by harmonised EU rules, including U-space and vertiport regulation, to enable consistent implementation as well as to facilitate cross-border operations. (European Commission, 2022a, 2022b).

REFERENCES

Advanced Technology Research Council. (2025, February 13). UAE begins mapping air corridors for air taxis and cargo drones to transform urban transportation. ATRC. https://www.atrc.gov.ae/news/uae-begins-mapping-air-corridors-air-taxis-and-cargo-drones-transform-urban-transportation

Alcock, C. (2025, July 15). *EASA issues innovative air mobility rules covering eVTOL aircraft. Aviation International News (AINonline).*

Amazon Staff. (2023, October 18). *Amazon is launching ultra-fast drone deliveries in Italy, the UK, and a third location in the U.S.* About Amazon.

Archer Aviation, Inc. (2023, March 23). United Airlines and Archer announce first commercial electric air taxi route in Chicago [Press release]. https://investors.archer.com/news/news-details/2023/United-Airlines-and-Archer-Announce-First-Commercial-Electric-Air-Taxi-Route-in-Chicago/default.aspx

Archer Aviation, Inc. (2024, December 6). Abu Dhabi and Archer announce agreement with cross-industry stakeholders to launch first commercial electric air taxi flights [Press release]. Archer Investor Relations. https://investors.archer.com/news/news-details/2024/Abu-Dhabi-and-Archer-Announce-Agreement-With-Cross-Industry-Stakeholders-To-Launch-First-Commercial-Electric-Air-Taxi-Flights/default.aspx

Archer Aviation, Inc. (2025, July 2). Archer begins test flights in Abu Dhabi [Press release]. Archer Investor Relations. https://investors.archer.com/news/news-details/2025/Archer-Begins-Test-Flights-in-Abu-Dhabi/default.aspx

AutoFlight. (2024a, February 27). AutoFlight showcases world's first inter-city air-taxi flight (Shenzhen–Zhuhai) [Press release]. https://www.autoflight.com/en/news/shenzhen-zhuhai-route/

AutoFlight. (2024b, April 3). *AutoFlight delivers first electric air taxi demonstrator to customer in Japan.* https://www.autoflight.com/en/news/autoflight-delivers-first-evtol/

Beijing Review. (2024, December 9). *Innovation drives a new horizon in the low-altitude economy.* https://www.bjreview.com/Business/202412/t20241209_800386517.html

Brinkmann, P. (2022, October 21). *An air taxi without a pilot aboard? This developer is seeking permission under existing regulations. Aerospace America (AIAA).*

Bridgelall, R., & Tolliver, D. (2024). Transforming healthcare delivery with advanced air mobility: A rural study with GIS-based optimization. *Sustainability, 16*(13), 5709. https://doi.org/10.3390/su16135709

Bristow Group Inc. (2025, March 4). *First formal agreement to use Norway as an international test arena for zero- and low-emission aviation signed* [Press release].

Butterworth-Hayes, P. (2019, March 16). Shenzhen hosts over 2,000 drone flights a day after UTM system launch. Unmanned Airspace. https://www.unmannedairspace.info/urban-air-mobility/shenzhen-hosts-over-2000-drone-flights-a-day-following-introduction-of-city-wide-utm-system/

Chin, C., Qin, V., Gopalakrishnan, K., & Balakrishnan, H. (2023). Traffic management protocols for advanced air mobility. *Frontiers in Future Transportation, 2*, 1176969. https://doi.org/10.3389/fpace.2023.1176969

Cirium Ascend Consultancy. (2025, January 14). *Advanced air mobility – Snapshot January 2025.* https://www.cirium.com/thoughtcloud/ascend-consultancy-advanced-air-mobility-snapshot-january-2025/

Cornell, A., Mahan, S., & Riedel, R. (2023, October 6). *Commercial drone deliveries are demonstrating continued momentum in 2023.* McKinsey & Company. https://www.mckinsey.com/industries/aerospace-and-defense/our-insights/future-air-mobility-blog/commercial-drone-deliveries-are-demonstrating-continued-momentum-in-2023

Daleo, J. (2024, August 12). Air taxis missed Paris Olympics goal—Could they soar in LA? *FLYING Magazine.* https://www.flyingmag.com/air-taxis-missed-paris-olympics-goal-could-they-soar-in-la/

Deng, I. (2024, November 6). *Shenzhen to invest US$1.7 billion in economy for flying cars, drones by 2026. South China Morning Post.*

Department for Business, Energy & Industrial Strategy, Department for Transport, & UK Research and Innovation. (2022, July 18). *New aerospace innovation to propel UK to growth and greener skies backed by £273 million* [Press release]. GOV.UK.

Designboom. (2024, August 8). Volocopter's eVTOL completes crewed test flight at a vertiport in Paris amid the 2024 Olympics. https://www.designboom.com/technology/volocopter-evtol-crewed-test-flight-vertiport-paris-2024-olympics/

Deutsche Post DHL Group. (2014, September 25). DHL Parcelcopter launches initial operations for research purposes [Press release]. Business Wire. https://www.businesswire.com/news/home/20140925005205/en/DHL-parcelcopter-launches-initial-operations-for-research-purposes

DroneDJ. (2024, August 13). Flytrex hits 100K drone delivery milestone in NC, Texas. DroneDJ. https://dronedj.com/2024/08/13/flytrex-drone-delivery/

DroneXL. (2025, March 30). Manna Drone Delivery claims profitability edge over industry giants. DroneXL. https://dronexl.co/2025/03/30/manna-drone-delivery-profitability-edge/

EHang. (2023, October 13). EHang successfully obtains type certificate for EH216-S passenger-carrying UAV system [Press release]. EHang. https://www.ehang.com/news/990.html

EHang. (2025a, March 31). EHang obtains first batch of air operator certificates for civil human-carrying pilotless aerial vehicles [Press release]. https://www.ehang.com/news/1196.html

EHang. (2025b, May 26). EHang reports first quarter 2025 unaudited financial results [Press release]. GlobeNewswire. https://www.globenewswire.com/news-release/2025/05/26/3087921/0/en/EHang-Reports-First-Quarter-2025-Unaudited-Financial-Results.html

EHang. (2025c, July 1). *EHang partners with Reignwood Aviation to create a new model of general aviation and eVTOL collaboration in the low-altitude economy.* https://www.ehang.com/news/1245.html

EUROCONTROL. (2021, January 28). *Greening the future: New CORUS-XUAM project will see EUROCONTROL coordinate a series of Europe-wide urban air mobility demonstrations.*

EUROCONTROL. (n.d.). *Concept of operations for European UTM systems – Extension for urban air mobility (CORUS-XUAM)* [Project page]. Retrieved January 23, 2026, from https://www.eurocontrol.int/project/corus-xuam

European Commission. (2020). *Air Mobility Urban – Large Experimental Demonstrations (AMU-LED)* (Grant agreement ID: 101017702) [Project fact sheet, CORDIS]. https://cordis.europa.eu/project/id/101017702

European Commission. (2022a, November 29). *A Drone Strategy 2.0 for a Smart and Sustainable Unmanned Aircraft Eco-System in Europe* (COM(2022) 652 final).

European Commission. (2022b, November 29). *EU Drone Sector state of play* (SWD(2022) 366 final).

European Union Aviation Safety Agency. (2021, May 19). *Study on the societal acceptance of Urban Air Mobility in Europe: Short report.* https://www.easa.europa.eu/sites/default/files/dfu/uam-short-report.pdf

European Union Aviation Safety Agency. (2025, July 18). *Special Condition for VTOL and Means of Compliance.*

Federal Aviation Administration. (2022). *Advanced Air Mobility (AAM) implementation plan (Version 1.0)* [PDF]. https://www.faa.gov/sites/faa.gov/files/AAM-I28-Implementation-Plan.pdf

Goyal, R., & Cohen, A. (2022). Advanced air mobility: Opportunities and challenges deploying eVTOLs for air ambulance service. *Applied Sciences*, *12*(3), 1183. https://doi.org/10.3390/app12031183

Goyal, R., Reiche, C., Fernando, C., & Cohen, A. (2021). Advanced air mobility: Demand analysis and market potential of the airport shuttle and air taxi markets. *Sustainability*, *13*(13), 7421. https://doi.org/10.3390/su13137421

Hamburg Aviation. (2025, February 12). Urban Air Mobility im Aufwind: Hamburg als Motor für die Luftfahrt_Revolution. Hamburg Aviation. https://www.hamburg-aviation.de/aktuelles/detail/urban-air-mobility-im-aufwind-hamburg-als-motor-fuer-die-luftfahrt-revolution/

Joby Aviation. (2024a, February 9). *Joby receives FAA approval for propulsion certification plan* [Press release].

Joby Aviation. (2024b, November 12). Joby announces beginning of work on first Dubai vertiport [Press release]. Joby Press. https://ir.jobyaviation.com/news-events/press-releases/detail/115/joby-announces-beginning-of-work-on-first-dubai-vertiport

Joby Aviation. (2025). Joby cements global lead in air taxi industry with Dubai flights and vertiports. Joby Press. https://ir.jobyaviation.com/news-events/press-releases/detail/132/joby-cements-global-lead-in-air-taxi-industry-with-dubai

Johnson, W., & Silva, C. (2021). NASA concept vehicles and the engineering of advanced air mobility aircraft. *The Aeronautical Journal, 126*(1299), 59–91. https://doi.org/10.1017/aer.2021.92

Kang, L. (2025, March 31). Chinese eVTOL maker EHang gets clearance for paid commercial operations. *CnEVPost*. https://cnevpost.com/2025/03/31/ehang-gets-operator-certificates/

Kiesewetter, L., Shakib, K. H., Singh, P., Rahman, M., Khandelwal, B., Kumar, S., & Shah, K. (2023). A holistic review of the current state of research on aircraft design concepts and considerations for advanced air mobility applications. *Progress in Aerospace Sciences, 142*, 100949. https://doi.org/10.1016/j.paerosci.2023.100949

Kwok, P. (2024, January 17). Low-altitude economy to fly high – China makes great strides in advanced air mobility development. Airports Council International Asia-Pacific & Middle East. https://www.aci-asiapac.aero/media-centre/perspectives/low-altitude-economy-to-fly-high-china-makes-great-strides-in-advanced-air-mobility-development

Lake Nona Editorial Team. (2020, November 11). *Lilium partners with Tavistock Development and City of Orlando to establish Florida as the first advanced aerial mobility region in the United States* [Press release]. Lake Nona.

Leedom, M. (2024, June 19). Drones deliver humanitarian aid in Africa. *Think Global Health (Council on Foreign Relations)*. https://www.thinkglobalhealth.org/article/drones-deliver-humanitarian-aid-africa

McCorkendale, Z., McCorkendale, L., Kidane, M., & Namuduri, K. (2024). Digital traffic lights: UAS collision avoidance strategy for advanced air mobility services. *Drones, 8*(10), 590. https://doi.org/10.3390/drones8100590

McKinsey & Company. (2023, October 6). Commercial drone deliveries are demonstrating continued momentum in 2023. McKinsey Future Air Mobility blog. https://www.mckinsey.com/industries/aerospace-and-defense/our-insights/future-air-mobility-blog/commercial-drone-deliveries-are-demonstrating-continued-momentum-in-2023

McKinsey & Company. (2024, October 1). Consumer views of drone delivery: How soon can you get here?. McKinsey Future Air Mobility blog. https://www.mckinsey.com/industries/aerospace-and-defense/our-insights/future-air-mobility-blog/consumer-views-of-drone-delivery-how-soon-can-you-get-here

McNab, R. (2023). Advanced air mobility, economic impacts, and equity considerations. *Journal of Economic Analysis, 3*(2), 155–170. https://doi.org/10.58567/jea03020009

Moradi, N., Wang, C., & Mafakheri, F. (2024). Urban air mobility for last-mile transportation: A review. *Vehicles, 6*(3), Article 66. https://doi.org/10.3390/vehicles6030066

Orlando Economic Partnership. (2020, November 11). *German-based startup Lilium to establish nation's first advanced aerial mobility region in Orlando* [Press release].

Polek, G. (2022, May 26). *Joby receives Part 135 operating certificate from the FAA. FutureFlight (Aviation International News).*

Pritchard, J. (2024, August 14). *UK: Advanced Mobility Ecosystem Consortium provides progress update on bringing AAM closer to reality.* eVTOL Insights. https://evtolinsights.com/uk-advanced-mobility-ecosystem-consortium-provides-progress-update-on-bringing-aam-closer-to-reality/

PwC. (2024, September 10). Drone deliveries: Taking retail and logistics to new heights. PwC Transportation & Logistics. https://cee.pwc.com/drone-powered-solutions/drone-deliveries-taking-retail-and-logistics-to-new-heights.html

Raghunatha, A., Thollander, P., & Barthel, S. (2023). Addressing the emergence of drones – A policy development framework for regional drone transportation systems. *Transportation Research Interdisciplinary Perspectives, 19*, 100795. https://doi.org/10.1016/j.trip.2023.100795

Reuters. (2025, February 13). UAE begins mapping corridors for air taxis and cargo drones. Reuters. https://www.reuters.com/world/middle-east/uae-begins-mapping-corridors-air-taxis-cargo-drones-2025-02-13/

Ros, M. (2024, April 4). *AutoFlight delivers first eVTOL to commercial customer in Japan. AeroTime.* https://www.aerotime.aero/articles/autoflight-delivers-first-evtol-to-commercial-customer-in-japan

SAFE (Coalition for Reimagined Mobility). (2025, February). *Beyond takeoff: Policy strategies to build the infrastructure future for advanced air mobility* [Report]. https://reimaginedmobility.org/wp-content/uploads/2025/02/ReMo_Beyond-Takeoff-Report_2025.pdf

Sampson, B. (2023, October 16). Chinese regulators certify EHang eVTOL. *Aerospace Testing International.*

Scalea, J. R., Restaino, S., Scassero, M., Bartlett, S. T., & Wereley, N. (2019). *The final frontier? Exploring organ transportation by drone. American Journal of Transplantation, 19*(3), 962–964. https://doi.org/10.1111/ajt.15113

Scott, B. I., & Öykü, K. (2025, August). *High hopes and higher hurdles: Unpacking six regulatory challenges facing Advanced Air Mobility,* Research in Aerospace Law & Policy Working Paper 2025/03 (Leiden University, International Institute of Air & Space Law). universiteitleiden.nl/binaries/content/assets/rechtsgeleerdheid/instituut-voor-publiekrecht/lucht–en-ruimterecht/working-papers/ralp2025-03.pdf

Shenzhen Daily. (2024, January 5). Regulations pave the way for soaring low-altitude economy. Nanshan Government Online. https://www.szns.gov.cn/english/news/latestnews/content/post_11088000.html

Skyports Infrastructure. (2024). Skyports and RTA break ground on first vertiport in Dubai (capacity up to 170,000 passengers/year) [Press release]. Skyports. https://skyports.net/skyports-and-rta-break-ground-on-first-vertiport-in-dubai/

Stonor, C. (2023, March 13). Wing drone delivery "capable of handling tens of millions of deliveries for millions of consumers" by mid-2024. *eVTOL Insights.* https://evtolinsights.com/wing-drone-delivery-capable-of-handling-tens-of-millions-of-deliveries-for-millions-of-consumers-during-2024/

Swiss Post. (2017, March 31). Swiss Post drone to fly laboratory samples for Ticino hospitals [Press release]. Swiss Post. https://www.post.ch/en/about-us/media/press-releases/2017/swiss-post-drone-to-fly-laboratory-samples-for-ticino-hospitals

Talaie, T., Niederhaus, S., Villalongas, E., & Scalea, J. (2021). *Innovating organ delivery to improve access to care: Surgeon perspectives on the current system and future use of unmanned aircrafts. BMJ Innovations, 7*(1), 157–163. https://doi.org/10.1136/bmjinnov-2020-000439

The Verge. (2025, July). Joby delivers first aircraft to Dubai as air taxi service nears launch. The Verge. https://www.theverge.com/news/694515/joby-delivers-first-aircraft-to-dubai-as-air-taxi-service-nears-launch

UK Research and Innovation. (2025, October 13). *Area of investment and support: Future flight programme.*

Urban Air Mobility News. (2024, July 20). Shenzhen announces plans for over 1,200 drone/eVTOL take-off and landing platforms. https://www.urbanairmobilitynews.com/air-traffic-management-u-space-uam-uam-infrastructure/shenzhen-announces-plans-for-over-1200-drone-evtol-take-off-and-landing-platforms/

Urban Air Mobility News. (2025, January 7). China's eVTOL sector accelerates into 2025 with new platforms, orders, and city plans. https://www.urbanairmobilitynews.com/new-city-projects/chinas-evtol-sector-accelerates-into-2025-with-new-platforms-orders-and-city-plans/

Wang, N., Mutzner, N., & Blanchet, K. (2023). Societal acceptance of urban drones: A scoping literature review. *Technology in Society, 75,* 102377. https://doi.org/10.1016/j.techsoc.2023.102377

Warwick, G. (2025, July 17). *EASA reinforces regulatory framework for eVTOLs. Aviation Week Network.*

Wheeler, P., Esselman, J., & Organ, A. (2024). Lessons learned while planning for advanced air mobility. *Journal of Airport Management, 18*(2), 142–155. https://doi.org/10.69554/CYSF8675

Wing. (2025). About Wing: Takeoffs, landings, and milestones. Wing.

Wing, D., Chancey, E., Politowicz, M., & Ballin, M. (2020, June). Achieving resilient in-flight performance for advanced air mobility through simplified vehicle operations. AIAA Aviation 2020 Forum. https://doi.org/10.2514/6.2020-2915

Zag Daily. (2025, January 9). Gary Vermaak's 2025 predictions for advanced air mobility. https://zagdaily.com/zag-air/gary-vermaaks-2025-predictions-for-advanced-air-mobility/

Zipline. (2024, April 19). Zipline makes one-millionth commercial autonomous drone delivery (and expands U.S. deliveries) [Press release]. Zipline Newsroom. https://www.zipline.com/newsroom/news/announcements/zipline-makes-one-millionth-commercial-autonomous-drone-delivery-and-expands-u-s-deliveries-to-include-panera-bread-memorial-hermann-health-system-and-jet-s-pizza

Part II

Resilient Safety

Advanced air mobility (AAM) represents a complex socio-technical ecosystem that encompasses new aircraft types, novel airspace integration, innovative operations, emerging infrastructure, and engaged communities. Ensuring safety in this dynamic environment requires more than traditional compliance checklists or reactive accident responses. Instead, it demands a resilient safety approach—a system-wide, proactive strategy that continuously adapts to changing conditions and uncertainties.

This section introduces the four pillars of resilience engineering—anticipate, monitor, respond, and learn—and illustrates how these capabilities apply across the five pillars of AAM (aircraft, airspace, operations, infrastructure, and community) to create a robust, adaptive safety system. In contrast to conventional safety models that focus on preventing specific failures or meeting minimum regulations, resilient safety emphasizes the capacity of the AAM system to perform safely under both expected and unexpected conditions. This aligns with the emerging shift in aviation safety regulation toward performance-based and risk-based frameworks (as evidenced by recent Federal Aviation Administration (FAA) and European Union Aviation Safety Agency (EASA) initiatives), which increasingly value continuous improvement and adaptability alongside compliance. For example, the FAA's 2024 powered-lift rulemaking established a flexible framework for pilot training and operations to enable early AAM services, emphasizing that regulations must "allow for innovation and new approaches" in certification.

THE FOUR PILLARS OF RESILIENCE ENGINEERING

Erik Hollnagel and colleagues characterize resilient system performance in terms of four key capabilities or "potentials," often assessed using a resilience assessment

DOI: 10.1201/9781003546306-5

grid. These four pillars—the abilities to anticipate, monitor, respond, and learn—provide a conceptual framework for how AAM organizations and systems can actively manage safety:

Anticipate: Knowing What to Expect

This is the ability to foresee potential future events, disruptions, or opportunities and prepare in advance. Anticipation involves identifying emerging risks (e.g., new failure modes, changes in weather patterns, or changes in traffic density) and envisioning novel scenarios. In practice, anticipation in AAM could include scenario planning for high-density vertiport operations or forecasting maintenance needs for a new electric vertical takeoff and landing (eVTOL) aircraft. It requires using leading indicators and predictive data to proactively address issues before they manifest, a key aspect of resilience engineering. High reliability organization (HRO) theory refers to this mindset as a "preoccupation with failure," meaning even minor anomalies are taken as signs to investigate and act upon (Weick & Sutcliffe, 2007). By anticipating, AAM stakeholders can implement design redundancies, contingency procedures, or regulatory guidance ahead of time to prevent accidents.

For example, regulators and industry might anticipate novel failure modes in autonomous AAM aircraft and develop standards for software assurance accordingly, rather than waiting for accidents to reveal those weaknesses.

Monitor: Knowing What to look for

This is the ability to attentively track both the internal state of the system and the external operating environment in real time. Monitoring means having situational awareness of factors that could affect safety in the near term—from vehicle health and battery performance to weather, traffic, and even community feedback. Effective monitoring in AAM uses sensors, data networks, and human oversight to detect early signs of trouble or changing conditions.

For instance, an AAM operations center might continuously monitor aircraft telemetry and airspace data to spot trajectory deviations or congestion developing. HROs exhibit a "sensitivity to operations," maintaining awareness of the front-line reality at all times (Weick & Sutcliffe, 2007). In the AAM context, this could include real-time tracking of vertiport status, communication links, and infrastructure health. The monitor capability ensures that the system knows what is normal versus abnormal, enabling prompt detection of deviations that require a response.

Respond: Knowing What to Do

This is the ability to act effectively and quickly in the face of regular or irregular disturbances. A resilient AAM system must be ready to respond to incidents, failures, or unexpected events—whether it is an in-flight emergency, a sudden weather change, or a cybersecurity breach. The response capacity involves having pre-prepared actions (e.g., emergency procedures, redundancies) and the agility to adjust operations on the fly.

An example in AAM would be an eVTOL aircraft's automated emergency landing system activating when a critical component fails, or an air traffic management system rerouting vehicles around a pop-up no-fly zone. Weick and Sutcliffe's (2007) research on HROs highlights a commitment to resilience—organizations devote resources to training and improvisation so that when surprises occur, they can contain and recover from them swiftly. In practical terms, this means AAM operators and pilots (or autonomous systems) are prepared for contingencies; they practice loss-of-control scenarios, have backup navigation methods in place if GPS is lost, and maintain the flexibility to adjust plans during disruptions. The ultimate goal of responding is to absorb shocks and maintain system functionality, or at least fail safely without catastrophic consequences.

Learn: Knowing What Has Happened

This is the ability to learn from past experience—to draw the right lessons from both successes and failures. Continuous learning closes the loop by feeding back insights into improved anticipation, monitoring, and response. In a resilient safety approach, every incident, near-miss, and even normal operation is an opportunity to improve the system. For AAM, this might involve analyzing data from thousands of drone or air taxi flights to refine operational protocols and design standards. It also means investigating accidents or anomalies not to assign blame (as in traditional *Safety-I* thinking) but to understand how the system can be made more robust (the *Safety-II* perspective). Dr. James Reason's work on organizational safety emphasized that many hazards stem from latent conditions—hidden system weaknesses that lie dormant until combined with active failures (Reason, 1997). A learning organization works to surface and address these latent conditions before they contribute to an accident.

In AAM, this could include updating software after discovering a hidden flaw in an autopilot's logic, or revising vertiport procedures after a near-miss on a congested pad. Learning is formalized through processes such as safety management systems (SMS), incident reporting systems, and data-sharing collaborations (e.g., the FAA's ASIAS or NASA's knowledge databases), ensuring that the entire community benefits from each lesson. As Reason famously noted, reliability in complex systems is essentially "a dynamic non-event"—constant adjustments prevent bad outcomes, so nothing bad happens. Effective risk management involves not only preventing errors but also building systems that can tolerate and recover from errors when they occur (Reason, 2000).

Importantly, these four resilience components reinforce each other as an integrated cycle. The act of responding to an event provides new information that can be learned from; that learning enables better anticipation of future conditions; anticipation focuses on what systems should monitor; and monitoring detects the events that require a response. A resilient AAM safety system is, therefore, dynamic and iterative, continually updating its expectations and strategies. This stands in contrast to older safety models that rely on static defenses alone (such as checklists or hard barriers). For example, whereas the classic *Swiss cheese model* of safety visualizes layered defenses passively catching errors (Reason, 1990), a resilience approach goes further by actively adapting those layers and human/automated responses in real-time as the system evolves. In the rapidly changing AAM environment, this adaptability is crucial.

Resilient safety in AAM is about building a safety net that is as agile and innovative as the technologies and operations it protects. By focusing on the how of resilient safety—how we anticipate the future, how we monitor the present, how we respond in the moment, and how we learn from the past—we create a living safety system that grows stronger with each experience. The four pillars of resilience engineering provide a practical lens for assessing and improving safety across all facets of AAM, from the design of an eVTOL aircraft to the community engagement surrounding a new vertiport. This system-wide, dynamic approach does not replace traditional safety measures; instead, it enhances them, ensuring that compliance is not the ceiling of safety performance, but the floor upon which continuous improvement is built.

As AAM moves from concept to reality, the industry and regulators are collectively navigating uncharted skies. Challenges will undoubtedly emerge—technical failures, unforeseen interactions, and human errors in new contexts. A resilient safety approach accepts this uncertainty not with fatalism, but with preparedness. It strives to make the AAM ecosystem robust yet flexible, able to withstand surprises and continue safe operation, or at least graceful recovery. In practical terms, this means investing in data capabilities, training, fostering strong safety cultures, and adapting to regulations. It means valuing the human expertise on the front lines as much as the automated systems, and viewing accidents not as endpoints of blame, but as lessons for the prevention of future mishaps. By embracing resilient safety, AAM stakeholders—engineers, operators, regulators, and communities—co-create an aviation system that can deliver on the promise of AAM safely, sustainably, and confidently. This proactive ethos of safety will help AAM earn public trust and achieve its vision of transforming transportation, while upholding the highest standards of safety that aviation demands.

4 Anticipate
Proactive Planning

INTRODUCTION

Advanced air mobility (AAM) represents a new paradigm in air transportation, encompassing highly automated electric vertical takeoff and landing (eVTOL) aircraft, novel airspace management concepts, new operational models, supportive infrastructure, and engaged communities. Ensuring resilient safety in this emerging domain requires a proactive approach. In resilience engineering, the first step is to anticipate—to foresee potential challenges and opportunities and prepare the system to handle them before they escalate. Anticipation is especially critical for AAM given its complexity and high stakes: vehicles will operate in urban skies, rely on automation, and involve diverse stakeholders.

Resilience engineering focuses on how systems can adapt to both seize opportunities and withstand unforeseen challenges. Instead of merely reacting to failures after the fact, a resilient AAM system must actively predict and prevent problems, creating safety through foresight and flexibility. Plans for AAM often envision high levels of automation to reduce human error, yet it is crucial to preserve human strengths in adaptation and problem-solving (Nemeth & Holbrook, 2021). The goal is to combine advanced automation with human insight to improve awareness, allocate resources wisely, and prepare for surprises. This chapter will examine how anticipation contributes to resilient safety across the five key layers of AAM introduced earlier:

- Aircraft—eVTOL vehicle design readiness, reliability, and emergent risk identification.
- Airspace—anticipatory traffic management and conflict avoidance strategies.
- Operations—mission planning with contingencies and organizational preparedness.
- Infrastructure—vertiport and support systems planning with foresight for capacity, weather, and hazards.
- Community—anticipating public perception issues, regulatory shifts, and societal impact.

THEORETICAL FOUNDATIONS OF ANTICIPATION

In resilience engineering, anticipation is recognized as one of the core capacities that a safe system must possess. Hollnagel (2009) defined *anticipating* as the capability to be prepared for developments further into the future, e.g., disruptions, constraints, or opportunities. It is one of the four basic potentials for resilient performance,

DOI: 10.1201/9781003546306-6

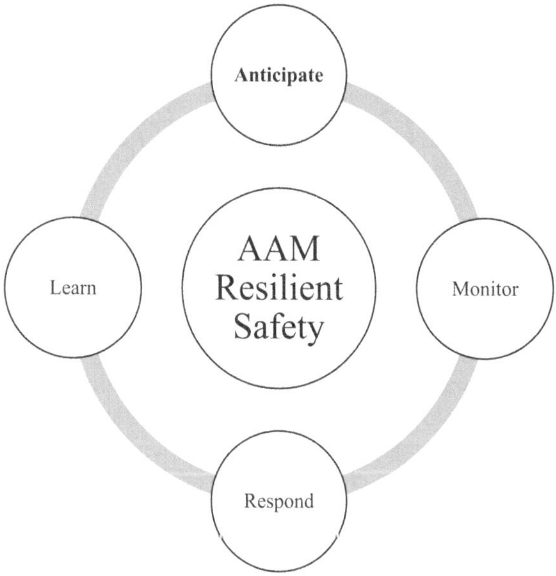

FIGURE 4.1 Anticipate function for resilient safety in AAM.

Note: Based on Hollnagel's (2009, 2015) four resilience abilities (anticipate, monitor, respond, learn).

alongside monitoring, responding, and learning. A resilient organization proactively imagines future events—both threats and opportunities—and readies itself accordingly. In practice, this means identifying emerging risks, spotting weak signals, and envisioning how situations could evolve beyond normal conditions. Hollnagel's resilience analysis grid (RAG) provides a framework to assess and enhance a system's ability to anticipate, by posing questions about how well an organization "knows what to expect" in the face of possible challenges. The potential to anticipate is essentially the ability to foresee future changes, novel demands, or constraints, and to prepare for them. For example, an AAM operator with high anticipatory capacity may conduct extensive scenario planning for rare events (such as GPS outages or cyberattacks on navigation systems) and develop contingency plans in advance. Hollnagel emphasizes that the four resilience potentials are interdependent—improving anticipation can aid responding and monitoring, and it relies on learning from past experiences.

Reason's work on human error and system safety also underscores the importance of anticipation, albeit in the more traditional Safety-I context. Reason's *Swiss Cheese Model* and his concepts of latent conditions highlight the need to anticipate problems in system design and organization. Latent conditions are the hidden problems in a system (poor design, training gaps, organizational deficiencies) that can combine with active failures (e.g., human errors) to produce an accident (Reason, 2000). Crucially, Reason noted that while specific human errors are hard to predict, latent conditions *can* be identified and addressed before an adverse event occurs. This turns safety management into a proactive endeavor rather than a reactive one. In his words, effective risk management involves "identifying and neutralizing" these latent failures before they combine to cause harm (Reason, 1997). Anticipation in this sense means systematically identifying and addressing weak points—the *resident pathogens* in the system—before they contribute to a mishap. For AAM, this could mean rigorously analyzing a new flight control algorithm for corner cases or auditing a maintenance program for gaps, given the novel electrical propulsion systems. It also means examining organizational factors, such as anticipating that pilots transitioning from helicopters to eVTOLs might misinterpret certain cues, and thus designing training to address this issue before any incident occurs.

Resilience engineering builds on these ideas, moving beyond preventing failure to also *promoting success*. While traditional Safety-I approaches (per Reason) focus on eliminating errors and adverse outcomes, resilience engineering and Safety-II thinking encourage us to also examine why things go right in everyday operations and how to anticipate variability (Hollnagel, 2018). This shift in perspective enables organizations to prepare for unexpected situations by fostering adaptation. As Hollnagel et al. (2006) noted, resilience engineering enables high-stakes sectors "to anticipate and sustain operations when confronted by unforeseen threats." In other words, it's not enough to react to yesterday's incidents; one must envision tomorrow's challenges. HRO theory similarly emphasizes anticipation. HROs (like aircraft carriers or nuclear power plants) are known for a "preoccupation with failure," meaning they constantly anticipate that something could go wrong and never assume an easy day. This mindset—described by Weick and Sutcliffe (2007)—leads to practices such as continual training for emergencies, aggressive near-miss reporting, and *what-if* brainstorming even when things seem to be going well. Although AAM is a relatively new field, it can adopt the principle of assuming that operations will be disrupted at some point and practicing adaptive responses in advance.

To make anticipation concrete, resilience practitioners use various tools. Scenario planning and contingency analysis allow teams to explore hypothetical events (e.g., a sudden communications blackout in an urban air taxi network) and develop strategies before they happen. Techniques like Hollnagel's functional resonance analysis method (FRAM) help map complex interactions and anticipate how small changes can reverberate through a system, see Figure 4.2.

Organizations can also self-assess their readiness by asking questions from the RAG: *Do we actively gather information about emerging risks? Do we imagine worst-case and best-case scenarios for our operations? Do we allocate resources (time, budget) for training and preparedness?* By answering these, gaps in anticipatory capacity can be identified.

For instance, a FRAM analysis of an "AAM airspace management" scenario in work as imagined (WAI) approach, Figure 4.3, could include functions performed by aircraft automation, the unmanned aircraft systems (UAS) traffic management (UTM) system, pilots, and air traffic controllers, and it might reveal that specific information dependencies (such as the timely sharing of UAS position data with ATC) are potential weak links.

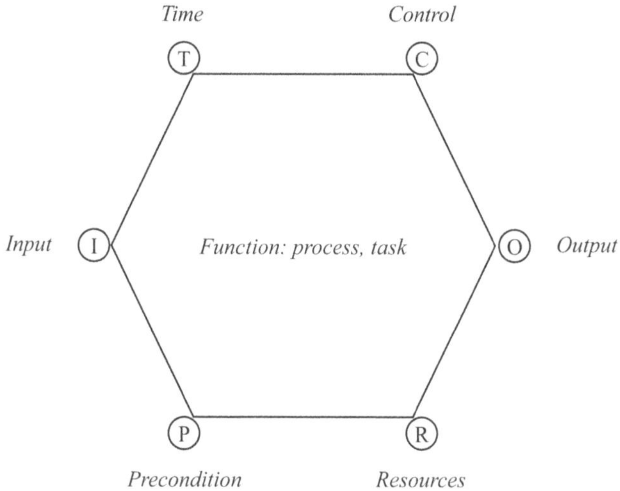

FIGURE 4.2 FRAM function aspects (Input, Output, Precondition, Resource, Control, Time).

Note. Adapted from Hollnagel and Hill's (2020) FRAM Model Visualiser (FMV) instructions (v2.1.4).

The WAI FRAM model would serve as a baseline for developing scenario-based risk assessments, which are confirmed via the work as done (WAD) FRAM model during the learning function of resilience engineering, covered in Chapter 7 (Figure 7.4), enabling continuous improvement within the organization's systems. The current WAI FRAM model describes the *intended* flow for deconflicting two AAM/UAM flights at a corridor intersection: a provider of services for UAM (PSU)/ UAS service supplier (USS) publishes the corridor and right-of-way cooperative operating practices (COPs); each operator submits a proposed operational intent (OI); the PSU detects a predicted conflict at the intersection and applies the COP to assign precise crossing slots; the updated intents are confirmed and shared across the federated USS/DSS (discovery and synchronization service) network; flights activate to meet their slots; conformance monitoring watches adherence in real time, and tactical detect and avoid (DAA) acts as a back-stop if needed; finally the operation is closed and lessons feed back into the COP. Overall, the model makes the dependencies explicit and sets the stage for the function description that could be monitored and then improved in the WAD FRAM process.

ANTICIPATION IN AAM AIRCRAFT SYSTEMS

The aircraft layer—particularly the design and operation of eVTOL vehicles and unmanned aircraft in AAM—must embody anticipatory safety from the outset. These aircraft are at the core of the ecosystem, and any failure in a vehicle can have immediate safety consequences. Anticipation at the aircraft level means

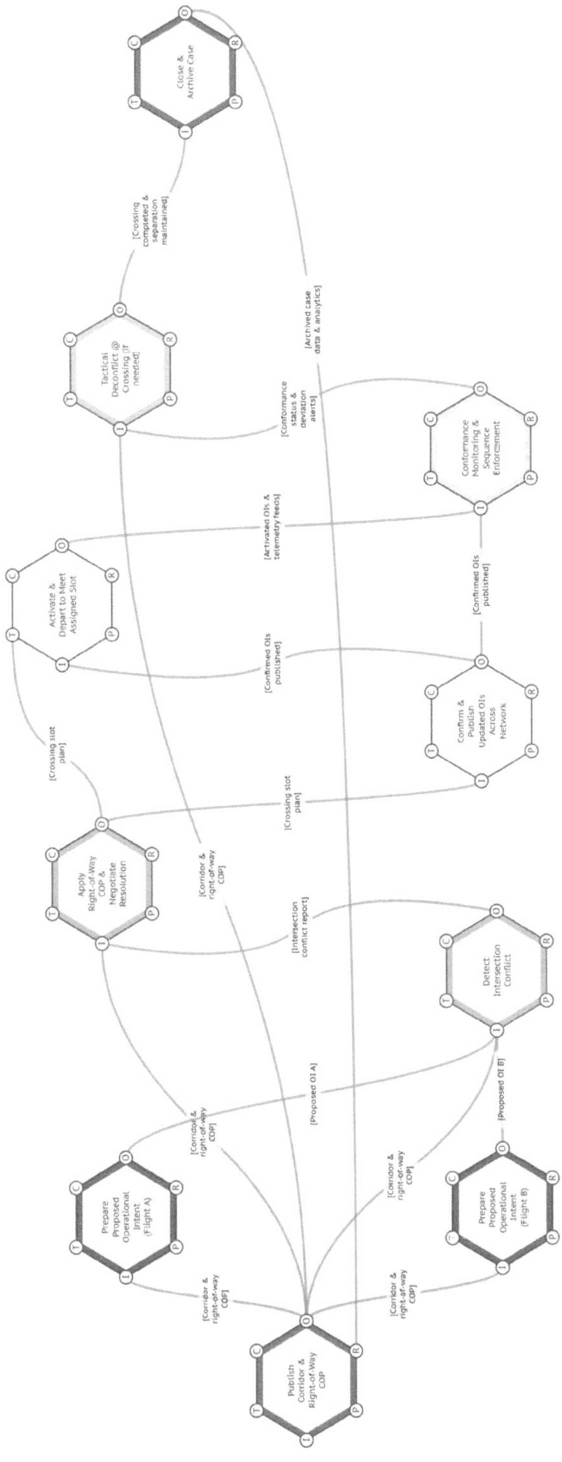

FIGURE 4.3 FRAM model for AAM airspace, work as imagined (WAI). (Credit: Author; Created with FRAM Model Visualizer.)

engineering the vehicle to handle a wide range of conditions and potential failures *gracefully*. Manufacturers and certifiers work to foresee what could go wrong and ensure the design either prevents those failures or tolerates them without catastrophe.

One major approach is anticipatory safety engineering, which includes designing in redundancies and fail-safes for critical systems well before any problem occurs. For example, many eVTOL aircraft have multiple independent lift rotors. Engineers anticipated that a propulsion unit could fail, so the aircraft is designed to remain controllable even if one (or more, in some designs) motors quit. The control system is programmed to automatically compensate for a lost rotor by increasing thrust on others—something that has been tested in simulations and flight tests ahead of time.

Similarly, flight control software goes through rigorous verification and validation under simulated abnormal conditions. Because eVTOLs are so software-dependent (for stability, navigation, autonomy, etc.), anticipating software failure modes or logic bugs is crucial. Developers use tools like fault injection testing—deliberately simulating sensor failures or corrupt data inputs in the flight software to see how the system reacts. This kind of testing, guided by imagination of what *could* happen, helps uncover latent software flaws or unhandled cases *before* the aircraft is in service. A relevant historical analogy is the development of the Airbus A320 in the 1980s (the first fly-by-wire airliner). Engineers had to anticipate how pilots might interact with flight control laws they'd never experienced, and they built in protections (like angle-of-attack and load factor limits) to prevent known hazards like stalls and overstresses. In AAM, the autonomy and novel propulsion bring new areas for anticipation—for instance, how will the aircraft behave if GPS signals are lost or spoofed? Anticipating that scenario leads to installing backup navigation systems (such as inertial systems or visual terrain mapping) so the vehicle can continue safely.

Maintenance and reliability engineering are also key areas of anticipation for aircraft. Rather than waiting for parts to break, AAM operators will employ predictive maintenance programs that utilize real-time data from aircraft systems to forecast when a component is likely to fail. For example, by continuously monitoring battery cell temperatures and charge-discharge cycles, algorithms can predict when a battery module is approaching the end of its life or is at risk of thermal issues, prompting replacement before a failure occurs. Similarly, tracking vibration signatures of motors or actuators can reveal early signs of wear. This data-driven anticipation of failures transforms maintenance from reactive to proactive, thereby reducing the likelihood of an in-flight failure. Modern airliners already utilize such systems, often referred to as health and usage monitoring systems, and eVTOLs are expected to follow suit. By identifying potential issues in advance (such as a rotor bearing starting to degrade), operators can service the aircraft on the ground rather than have the bearing seize in flight. This improves safety and also operational uptime.

An illustrative example of why anticipation is critical in aircraft systems comes from the analysis of automation surprises in aviation. The crash of Air France 447 in 2009—a perfectly airworthy Airbus A330 that lost control because the pilots became disoriented after the autopilot disengaged—revealed a failure of anticipation in both design and training. Neither the designers nor the operators had fully anticipated the scenario of high-altitude pilot icing leading to autopilot shutoff and pilot confusion. Post-incident, Airbus and airlines revamped procedures and training to anticipate and practice precisely that scenario (high-altitude manual recovery).

For AAM, which will rely even more on automation and even autonomous operations without pilots on board in some concepts, such lessons weigh heavily. Manufacturers are keenly aware that automation can create new failure modes if not anticipated. For example, an autonomous drone might encounter a situation its AI has not encountered, such as a GPS/compass discrepancy in a tight urban canyon, and behave unpredictably. To anticipate this, designers are incorporating heuristic "safety nets" into automation. If the system becomes uncertain or senses a condition outside its knowledge, it will default to a conservative mode, such as hovering in place or using a predefined safe landing routine, rather than proceeding. They are also extensively simulating edge cases in high-fidelity environments (including digital twins of cities and thousands of randomized scenarios) to expand what the AI *has* seen before it ever carries passengers.

Certification authorities are adapting, too. EASA and FAA are working on special conditions and means of compliance for autonomous systems and AI in aviation, essentially anticipating that conventional certification must evolve to address learning algorithms and complex software. This regulatory anticipation ensures that, by the time fully autonomous AAM vehicles seek approval, criteria are in place to evaluate their safety performance in all foreseeable situations, as well as some unforeseeable ones, through requirements for resilience and fail-safe behavior.

Figure 4.4 builds on Nemeth and Holbrook (2021), noting humans must remain "*the most capable source of information ingestion, situation understanding, and real-time decision adaptation*" even as automation expands. The graph presents the *Human–Automation Teaming Curve*, which models how *human cognitive contribution* (e.g., adaptation and anticipation) varies with *automation level*, peaking at

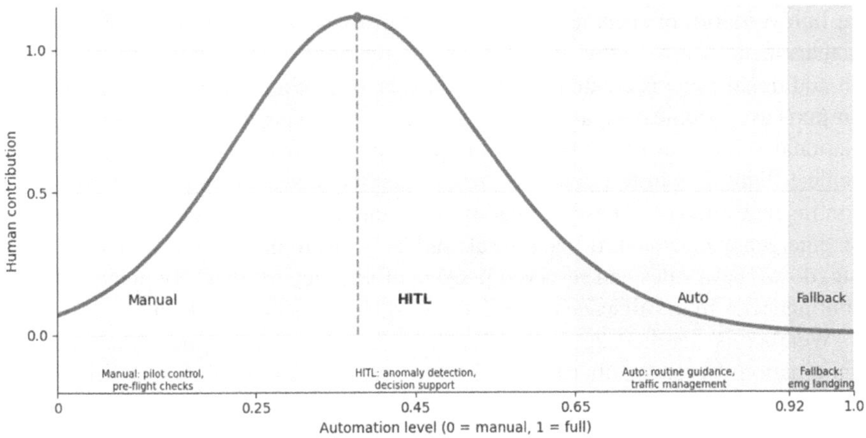

FIGURE 4.4 Human-automation teaming: preserving human oversight in anticipation roles. (Credit: Author, Data source: Chiou & Lee, 2021; Endsley, 1995; Hollnagel, 2017; Lee & See, 2004; Nemeth & Holbrook, 2021; Parasuraman et al., 2000; Prinzel & Ellis, 2021; Wickens et al., 2004.)

intermediate automation where human oversight is most critical. This framework is rooted in resilience engineering's focus on preserving human strengths in highly automated systems, ensuring operators remain engaged to detect anomalies and seize opportunities. Trust in automation research highlights that both overreliance and underreliance stem from poorly calibrated trust, reinforcing the need for *human-in-the-loop* zones where human judgment complements automation. Design principles for graceful degradation mandate that automation must relinquish control smoothly, enabling fallback to full human intervention when required—captured in the curve's *Fallback* phase.

ANTICIPATION IN AAM AIRSPACE MANAGEMENT

Managing the airspace for AAM operations requires significant anticipation, because we are introducing a large number of new aircraft—many of them small, automated, and operating in urban low-altitude airspace—into an already complex aviation system. The concept of UTM is still evolving, and to ensure safety, airspace managers must proactively foresee traffic scenarios, conflict hotspots, and system failure modes *before* AAM traffic densities grow.

One key area of anticipation is developing scalable airspace design and traffic flow concepts. Traditional air traffic control is largely tactical, handling individual flights with separation rules and human controllers. That model will not scale to hundreds or thousands of AAM flights over a city per day. Therefore, researchers and ANSPs anticipate high-density operations by proposing strategic deconfliction, such as designated AAM routes or corridors, altitude stratification, where different layers are allocated for different directions or vehicle types, and scheduling mechanisms to meter traffic. By simulating future peak traffic (e.g., a rush hour scenario with 200 air taxis and 100 delivery drones operating simultaneously in downtown), they identify potential choke points. Perhaps all flights tend to converge over a particular hub vertiport, or specific routes intersect in ways that could overwhelm conflict management. Knowing this ahead of time, mitigations can be put in place—perhaps an additional route is created to balance loads, or a requirement is established that air taxis use specific navigation performance to stay on precise paths. NASA's AAM National Campaign has conducted airspace simulations that integrate real AAM vehicle flight data to test concepts such as corridors, sequencing at vertiports, and contingency airspace. These simulations help anticipate issues such as how to handle an emergency diversion: if one aircraft suddenly has to make a precautionary landing, do we have rules and reserved pockets of airspace for that? By planning these contingencies in the airspace design, we avoid chaotic situations later.

Weather is another significant factor in airspace management that requires anticipation. Urban weather (microclimates, rapid storm development, wind gusts between buildings) can change quickly and impact AAM vehicles that are more susceptible to winds or icing than larger aircraft. Anticipation means integrating high-resolution weather forecasting and real-time weather data into the traffic management decision process. For instance, if convective cells are forecasted to develop along a commonly used corridor within the next 30 minutes, the system might stop assigning new flights to that corridor and reroute pending flights elsewhere before the storms arrive. Currently, metro helicopter services are sometimes surprised by pop-up thunderstorms; an anticipatory AAM system would leverage advances in nowcasting to reduce these surprises. This could involve vehicles themselves acting as weather sensors (a network of eVTOLs feeding back low-altitude wind and temperature data) to improve short-term forecasts for those altitudes. In this way, each AAM vehicle contributes to anticipating environmental challenges for the others—a concept sometimes referred to as a *network weather watch*.

Technology and infrastructure failures also need anticipation in airspace management. For example, loss of GPS is a well-known risk, whether due to interference or outages. A resilient approach is already anticipating this by developing backup navigation solutions for AAM, such as ground-based augmentation signals, vision-based navigation, or multilateration systems in urban areas. Some cities are even exploring terrestrial positioning networks as a contingency for GPS in drone corridors. Another example is communication links. AAM flights will likely depend on digital data links to UTM services or each other. Anticipating outages or congestion of these links, designers are looking at multi-link strategies (vehicles might have cellular 5G as primary and satellite or mesh network as secondary). Airspace procedures are being conceived for *lost link* scenarios. For example, if an air taxi loses contact with the network, it may have a predetermined protocol, i.e., continue to an emergency landing site or fly a holding pattern that everyone is aware of. By imagining these failure scenarios now, airspace rules can be written and vehicles designed to handle them with minimal risk.

On the regulatory side, bodies such as ICAO and EUROCONTROL are actively anticipating the challenges of AAM integration. As noted in the Eurocontrol *Unified Sky* initiative, there is an urgent push to update global ATM concepts and frameworks to include AAM. This includes establishing global contingencies and standard vocabularies—for instance, standardized priority rules if an AAM aircraft needs to enter controlled airspace for an emergency, or how to classify and handle unmanned AAM flights crossing international borders. Thinking through these issues *before* they happen (such as an air taxi unexpectedly entering a major airport's airspace to avoid weather) allows for creating coordination procedures and training controllers in advance.

Figure 4.5, inspired by the NASA UTM conflict mitigation model (NASA ARMD, 2020; n.d.), represents the aspect of anticipation in the AAM ecosystem. This chart visualizes how the required conflict-detection lead time must increase as UTM traffic density grows, illustrating four-layered mitigation thresholds drawn from NASA's UTM Technical Capability Level 4 conflict-management framework and ConOps guidelines. It demonstrates the gap between bare-minimum algorithmic alerts and the time-staggered buffers needed for geofencing, human override, and fallback procedures under high-density conditions, and models how the required conflict detection lead time must increase with higher UTM traffic densities. Therefore, mitigation layers could be introduced, such as algorithmic minimum alert, dynamic geofence buffer, human override margin, and fallback breakdown tolerance, highlighting the gap between minimal detection capabilities and the operational lead times needed for human or fallback interventions.

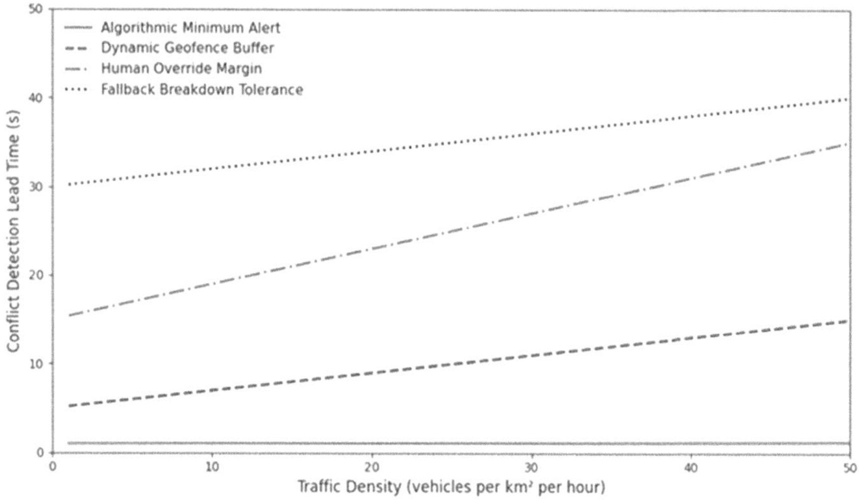

FIGURE 4.5 Traffic density vs. conflict prediction lead time in UTM traffic scenarios. (Credit: Author, Data source: Johnson & Larrow, 2020; University of Utah Kahlert School of Computing, 2018; NASA ARMD, n.d.; Jung & Craven, 2020; Rios et al., 2020; ICAO, 2005; ICAO, 2016; NASA ARMD, 2020; Chakrabarty et al., 2019; Johnson, 2020; Guerreiro et al., 2023.)

ANTICIPATION IN AAM OPERATIONS AND MISSION PLANNING

Beyond designing safe vehicles and airspace, AAM service providers must anticipate challenges in day-to-day operations and mission planning. This is where the "rubber meets the road" (or rather, the rotor meets the air): each flight must be planned and conducted with foresight to prevent minor issues from escalating into major problems. Anticipation in operations ranges from individual flight planning to organizational strategies and training programs.

During flight planning, AAM operators should consider not only normal operations but also credible contingencies and emergency procedures. A safety operational risk assessment framework in EASA's (2025) AMC/GM to Regulation (EU) 2019/947 (SORA) provides a practical structure for defining the relevant operational areas, including the flight geography (FG) for normal procedures, the contingency volume (CV) for abnormal situations and recovery actions, the operational volume (OV=FG+CV) used for containment and risk determination, the air risk buffer (ARB), and the ground risk buffer (GRB), as depicted in Figures 4.6 and 4.7. Considering such areas will allow normal operations while a UAS is flying inside the FG, contingency operations such as return to home (RTH), return to land (RTL), orbit, or manual takeover while crossing the outside boundaries of the FG and entering the OV, and then switch into emergency operations as soon as it crosses outbound the outside boundaries of the CV leading to automatic activation of the final termination system (FTS) and ultimately falling to the ground in a ballistic trajectory within the boundaries of the GRB (Scenario A in Figure 4.6). Suppose the UAS encounters a significant issue that leads to a continuous, uncontrollable climb beyond the vertical

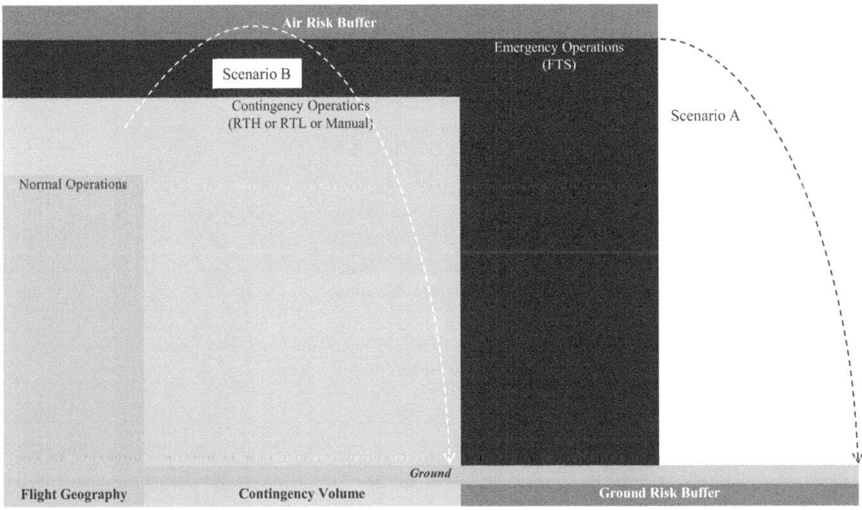

FIGURE 4.6 Operational areas, side view.

limits of the FG and the CV. In that case, the FTS will be automatically activated again to prevent the UAS from flying beyond the boundaries of the ARB, as depicted in Scenario B in Figure 4.6. In both cases, it is the planners' and operators' responsibility to ensure that flight planning is accurate, that all operational areas' values are accurate, that no people or property on the ground or aircraft in the air are affected, and that emergency procedures are in place.

Furthermore, a UAS operator shall embrace tools such as pre-flight risk assessments for every mission. For example, before a drone delivery flight or air taxi trip, dispatchers may go through a checklist or use a software tool that evaluates factors such as weather, wind, solar storm activity (for GPS), vertiport status, alternative landing options, and known airspace constraints. If any factor is marginal, they anticipate potential outcomes. *If* the wind picks up beyond limits at the destination, what is Plan B? Maybe Plan B is an alternate vertiport that's pre-checked for availability. *If* a sudden TFR (temporary flight restriction) is published because of, for example, a VIP movement or a sporting event, is there a contingency route filed? By addressing these *what-if* questions during planning, the operation can pivot more smoothly when unexpected events occur. This is analogous to how commercial airlines file alternate airports and carry extra fuel; an AAM flight might file an alternate vertiport and carry extra battery reserve. The need is perhaps even greater for AAM because the vehicles have less redundancy (one pilot for multiple flights, limited energy margins) and operate in more variable low-altitude environments. An example of hazard identification and risk assessment (HIRA) for AAM operations is listed in Annex C (European Union Aviation Safety Agency, n.d.; UK Civil Aviation Authority, 2024).

Operators are also using simulation and scheduling models to anticipate daily operations challenges. For instance, how to schedule charging at vertiports so that not all vehicles need a charger at once. By simulating a day of operation with dozens

FIGURE 4.7 Operational areas, top view.

of flights, an operator might discover that if three flights are delayed, it could lead to a queue at a charging pad that cascades into many other delays. Knowing this, they could proactively stagger schedules or secure additional charging capacity. They can also plan maintenance rotations by predicting when each vehicle is likely to reach specific usage or wear thresholds, thereby smoothing out the workflow instead of waiting for issues to arise. Essentially, data analytics and predictive algorithms are used to *anticipate demand and resource needs.*

Crew management is another area of anticipation. For piloted AAM operations, operators must anticipate potential issues such as pilot fatigue or excessive workload. They might restrict the number of high-workload flights, such as multiple take-offs and landings per hour in busy airspace, that a pilot can perform consecutively. Even for remote operators overseeing autonomous flights, anticipating human limitations is vital—for example, how many drones can a single supervisor realistically monitor? NASA and others are researching this to define safe ratios, which is a form of anticipating human performance limits before incidents reveal them the hard way.

From an organizational standpoint, AAM companies are trying to build a culture of anticipation. One way is through recurrent training and scenario drills. For example, Joby Aviation has reported that it is training its remote test pilots on how to handle various contingencies and developing checklists for scenarios like *GPS signal degraded* or *communication loss*. By doing this now, they anticipate that such events will happen eventually in operations, and their staff will be ready. Some operators run "day-in-the-life" simulations, where multiple teams simulate an entire operational day with various disruptions introduced (such as bad weather, unexpected VIP flight closures, vertiport power loss, etc.). These exercises function like wargaming—they reveal unanticipated interactions and help the team learn to adapt.

At the broader enterprise level, operators are anticipating how to scale safely and effectively. They know that a handful of vehicles can be managed somewhat manually, but for dozens or hundreds, automation in fleet management and SMS will be crucial. So, companies are investing early in operations management software that tracks every flight, every maintenance action, pilot scheduling, and weather feeds, providing a centralized, big-picture view. This not only facilitates real-time monitoring but also enables forecasting—for example, the system might alert that tomorrow's weather looks marginal for certain vertiports, prompting proactive cancellations or route changes the night before, thereby anticipating disruptions to minimize last-minute chaos.

In essence, anticipation in AAM operations is about being ahead of the game. It is the mindset that "forewarned is forearmed." By deeply integrating risk assessment into planning, using simulations to work through possible futures, and cultivating an HRO-like vigilance even on routine days, AAM operators aim to prevent incidents and ensure smooth service. When done well, much of this anticipatory work will be invisible to customers—flights will simply seem to run routinely. But behind the scenes, it will be the result of careful planning: nothing *bad* happened because the team anticipated it and mitigated the risk beforehand.

ANTICIPATION IN AAM INFRASTRUCTURE AND ENVIRONMENT

The *infrastructure* supporting AAM includes vertiports (for takeoff, landing, and charging), communication and navigation networks, and other ground facilities, such as maintenance hubs. Anticipating needs and risks in infrastructure is crucial because these elements can often become limiting factors or single points of failure if not adequately managed. AAM infrastructure must be robust against both physical and digital threats and scalable to meet future demand, which requires forward-thinking design and planning.

Starting with vertiports, city planners and companies are anticipating future demand and operational scenarios in how they design vertiports today. For instance, even if a vertiport initially handles 10 flights per hour, designers envision a future with 50 flights per hour and ensure the layout can accommodate this through modular expansion, additional pads, or rapid charging systems. They are also considering site-specific hazards. Take wind: rooftop vertiports in urban areas may experience strong downdrafts or turbulence. Rather than discovering after opening that approach paths are too turbulent, wind tunnel testing and computational fluid dynamics simulations are used during design to anticipate these effects and perhaps add baffles or specify approach angles that avoid the worst turbulence. Similarly, they anticipate the need for weather resilience: vertiports in cold climates are planned with de-icing equipment and heated pads, knowing eVTOL rotors could accumulate ice. In contrast, those in hot climates consider shaded parking and cooling for batteries.

One anticipatory design element in some proposed vertiports is redundancy in power. Because many eVTOLs will charge there, a vertiport might have both a connection to the city grid and on-site solar with battery storage or a generator. This way, if the grid goes down, which has happened at major airports, causing chaos, AAM operations could continue at least for a while on backup power—an investment made because they anticipated that power outages, while rare, could be disruptive and a safety risk, such as dark vertiport, no charging, etc. Indeed, recent events like the 2021 Texas power crisis or the 2019 Newark Airport blackout are reminders that infrastructure failures happen, and resilient design mitigates them.

On the digital infrastructure side, anticipation is about ensuring robust communication and navigation. AAM vehicles might rely on cellular networks to receive traffic info or flight clearances. Anticipating that cell networks can become overloaded or experience dead zones, some AAM trials are incorporating alternative communication methods, such as dedicated aviation frequencies or satellite links, for backup. Similarly, GPS is vital for navigation, but its vulnerabilities, i.e., jamming, urban canyon multipath, are well known. To address this, providers are testing augmentation systems, such as ground-based beacons or the European Galileo, as complementary solutions.

Another aspect is cybersecurity. Although there has not yet been a major malicious hack on a drone operation, the industry is anticipating this threat. Thus, from the ground up, AAM communication protocols are being designed with encryption and authentication in mind. EASA, for example, has included requirements for security management in its U-space regulations; this anticipatory step aims to prevent incidents such as spoofed commands or hijacked navigation that could lead to accidents or misuse of AAM vehicles.

Infrastructure anticipation also involves capacity management. For example, before a city approves five new vertiports, it wants to ensure that the local airspace can handle the traffic and that ground transportation can handle the throughput of passengers. There is a need to anticipate intermodal connections: if an air taxi drops someone at a vertiport, how will they complete their journey? City planners are thus incorporating AAM vertiport plans into urban mobility studies to anticipate issues such as congestion on surrounding roads or the need for public transit links. This might not seem like a direct safety issue. However, if a vertiport's pickup and drop-off zones are congested with cars, an AAM operator might rush turnarounds or have frustrated passengers, indirect pressures that could degrade safety margins. By anticipating and planning vertiport access and egress smoothly, these human-factor risks could be mitigated.

Environmental and community factors also fall under infrastructure anticipation. Vertiports will produce noise and change city skylines. Proactively measuring and modeling noise impact (using tools that simulate eVTOL sound propagation) enables the placement of vertiports in less sensitive areas or the addition of sound barriers. Planners anticipate that specific sites may raise community concerns, such as a vertiport near a park, so that they might adjust operational hours or flight paths during the design phase. Additionally, considering environmental resilience, if a vertiport is located in an area prone to flooding, raising it or waterproofing critical equipment is a proactive measure. The same applies to designing drainage systems for heavy rain, lightning protection, and other similar considerations. We see parallels in heliport design—experienced operators now build with these considerations in mind due to lessons learned—but AAM must accelerate this learning by anticipating it upfront.

ANTICIPATION IN COMMUNITY AND REGULATORY CONTEXT

Finally, we consider anticipation in the community and regulatory context of AAM. As discussed earlier, the community includes the public, local authorities, and other societal stakeholders who can significantly influence AAM's success or failure. Regulatory context refers to the evolving rules that govern AAM. Anticipating in this layer means looking ahead to potential public concerns and regulatory changes, and preparing accordingly. This is somewhat less tangible than engineering or operations anticipation. However, no less important, many aviation innovations have faltered not due to a lack of technology, but rather because of public or regulatory pushback that could have been avoided with proactive engagement.

CASE STUDY: LOSS OF GPS AND MANUAL REVERSION—SINGAPORE UAS INCIDENT

BACKGROUND

In February 2024, an incident in Singapore emphasized the importance of anticipating system failures and ensuring humans are prepared to take over. The scenario involved a Beyond Visual Line of Sight (BVLOS) drone demonstration flight conducted by Skyports Drone Services near Sentosa Island, Singapore (Transport Safety Investigation Bureau [TSIB], 2024). The drone was a ST Engineering Aerospace DrN-355LS, a six-rotor UAV weighing approximately 36 kg, which was flown autonomously along a predetermined route over water. The remote pilot (or UAS operator) was monitoring the operation from a Ground Control Station on shore, and the UAV was equipped with dual GNSS receivers and a fail-safe mode that allowed it to hover and permit manual takeover in the event of an automation issue. Both the technology and the pilot's training were supposed to handle a loss of GPS/GNSS signal—a known risk in urban or complex electromagnetic environments (interference, jamming, or signal loss can occur). However, as this case shows, having a backup system on paper is not enough unless operators are truly ready to use it under pressure.

Chronology of Events

Seven minutes into the flight, the drone's flight computer detected an anomaly: a "POSITION X-CHECK FAIL" error appeared, indicating a large discrepancy between the primary and secondary GPS readings. In essence, the UAV's secondary GNSS receiver had momentarily fed an invalid position (an impossible latitude coordinate), creating a position difference of nearly 10 km between the two receivers. The drone's software, unable to reconcile this, automatically switched from waypoint navigation to fail-safe hover mode. At that point, an aural alarm sounded, and a message flashed on the control screen, indicating that GPS was unreliable and prompting the pilot to take manual control. The system was designed to hover for 5 seconds, waiting for the pilot's input. If the pilot failed to take over in that window, the UAS would initiate an autonomous emergency landing (in this case, over water). Unfortunately, in the 2024 incident, the remote pilot did not regain control within the 5-second window, so the fail-safe logic kicked in: the drone began descending. The pilot still had a last-chance option to regain control during descent by issuing a Hover–Manual command via the GCS screen (see Figure 4.8), but the time available depended on altitude (at only ~50 m height, the window was very short), while his first reaction was to take control via the handheld controller (in contrary to the manufacturer's guidelines). Unfortunately, the UAS pilot never managed to reassert control; the UAV descended into the sea, *ditched*, and sank. It was a total loss of the aircraft, though mercifully no one was injured since it fell into water away from people.

Post-incident investigation revealed the human factors at play. The 44-year-old remote pilot was licensed and had nearly 99 hours of experience with UAVs, including

FIGURE 4.8 "Hover–Manual" command button on the GCS screen. (*Source*: TSIB, 2024.)

training on this specific drone type. He had been trained on the "*Position X-check Fail*" scenario—the prescribed response was to switch to hover-manual mode immediately. Yet in the heat of the moment, he froze. He later described being "*startled*" by the sudden failure: by the time he recalled the correct procedure, the UAV was already sinking. This highlights a classic problem in safety: a backup system (and procedure) that *theoretically* works, but in practice, the operator was unable to execute it in time. The incident investigation, conducted by the Singapore Transport Safety Investigation Bureau, cited both the technical trigger (the GNSS anomaly) and the operational shortfall (the pilot's delayed response). It issued recommendations to address these gaps. Specifically, the inquiry recommended improvements to pilot training for handling such emergencies and better design of the manual reversion process—either extending the available takeover time or making the cues and controls more user-friendly. For example, one suggestion was to make the emergency procedures more readily accessible to the pilot (the drone's manufacturer had not displayed the complete checklist of actions on the screen to avoid clutter, but the operator had created a separate *quick reference card*, which the pilot did not consult in time). Another idea was an audible alert or automated prompt that immediately instructs the pilot to take over controls, thereby reducing reaction time. In essence, they urged the operator to conduct more frequent drills and refresher training for pilots on time-critical failures, and for the manufacturer to improve the human factors design for alerts. We can infer that if the pilot had recognized and acted on the GPS failure within a few seconds, the drone could have been saved by switching fully to manual control.

Analysis and Lessons for AAM

For AAM operators and eVTOL developers, the lesson is clear: *anticipate that primary systems may drop out, and be fully prepared for it.* In traditional aviation, redundancy and failure management are core to safety (e.g., airliners routinely practice engine-out procedures in simulators). Similarly, AAM aircraft will rely heavily on systems like GPS for navigation, but GPS/GNSS is vulnerable to outages, interference, or spoofing. A resilient AAM operation must assume that such a loss *will* happen at some point and design both the technology and the human response to handle it gracefully. This means two things: (1) robust fallback systems on the aircraft, and (2) rigorous training/automation for pilots or remote supervisors to manage the transition.

On the technology side, AAM vehicles (including drones and piloted eVTOLs) are increasingly incorporating multi-sensor navigation systems. For example, modern UAVs can use inertial measurement units, vision-based navigation, and other sensors to maintain stable flight even if GPS is lost or jammed. One avionics developer recently unveiled a visual-aided navigation system that allows drones to continue accurate positioning in "GPS-denied" environments by combining camera imagery with inertial navigation. Many eVTOL prototypes plan to integrate GPS with radar/laser altimeters, optical sensors, and map data to provide "alternative positioning" in the event of satellite signal failure. Regulators also often require a *lost link* safety mode for unmanned aircraft, typically an automated hover or RTH feature if communications are lost. The Singapore drone *did* have such a mode (auto-hover then auto-land), which is beneficial, but the incident reveals that the mode's parameters (5-second timeout) and the pilot's readiness were insufficient.

Thus, the human and operational sides are just as crucial. Training remote pilots or AAM pilots to handle automation surprises must be a priority. The Singapore case recommended increasing the frequency of emergency drills and ensuring pilots do not have to rely solely on memory in high-stress moments. For AAM, this could translate to requirements that flight crews (or UAS operators) regularly practice sudden transitions from automated to manual flight in simulators or during training flights. They should be intimately familiar with the user interface and be able to execute critical actions (like engaging a backup control mode) in seconds. It may also be wise to design AAM cockpits/ground stations with more intuitive alerts—e.g., a distinct auditory alarm and flashing instruction like "GPS SIGNAL LOST – TAKE MANUAL CONTROL NOW" to cut through the confusion and prompt immediate action.

Finally, AAM operators should incorporate procedural anticipation: assume that at some point, Plan A (primary system) will fail, so always have Plan B ready. In crewed eVTOL operations, Plan B may involve the pilot reverting to manual flying using backup instruments. In remotely supervised or autonomous operations, Plan B might be an automatic safe landing or return to base if control is lost. The key is that these backups are in place from day one of operations, not added reactively after an incident. The Singapore UAV incident, while not catastrophic, serves as a valuable warning. A GNSS outage or spoofing event is a foreseeable scenario that *should not* cause an accident if well anticipated and rehearsed. By learning from it, AAM stakeholders can ensure that loss of a primary navigation signal becomes a non-event: the vehicle seamlessly switches to other navigation sources or awaits pilot command, and the operator responds immediately and correctly. In sum, redundancy plus rehearsal is the formula—build reliable fallback systems and train people to use them without hesitation.

REFERENCES

Chakrabarty, A., Stepanyan, V., Krishnakumar, K., & Ippolito, C. A. (2019). Real-time path planning for multi-copters flying in UTM-TCL4. AIAA Scitech 2019 Forum. https://doi.org/10.2514/6.2019-0958

Chiou, E. K., & Lee, J. D. (2021). Trusting automation: Designing for responsivity and Resilience. *Human Factors: The Journal of the Human Factors and Ergonomics Society*, *65*(1), 137–165. https://doi.org/10.1177/00187208211009995

Endsley, M. R. (1995). Toward a theory of situation awareness in dynamic systems. *Human Factors*, *37*(1), 32–64. https://doi.org/10.1518/001872095779049543

European Union Aviation Safety Agency. (n.d.). *Predefined risk assessment (PDRA)*. Retrieved November 1, 2025, from https://www.easa.europa.eu/en/domains/drones-air-mobility/operating-drone/specific-category-civil-drones/predefined-risk-assessment-pdra

European Union Aviation Safety Agency. (2025, September 29). *ED Decision 2025/018/R: Regular update of the AMC and GM to Commission Implementing Regulation (EU) 2019/947 — Operation of drones in the 'open' and 'specific' category (introducing SORA 2.5 in AMC1 to Article 11).*

Guerreiro, N. M., Hagen, G. E., Maddalon, J. M., & Butler, R. W. (2023). Analysis of strategic conflict management approaches as applied to simulated UAM operations (NASA/TM–20220017551). NASA. https://ntrs.nasa.gov/api/citations/20220017551/downloads/NASA-TM-20220017551.pdf
Hollnagel, E. (2009). The four cornerstones of resilience engineering. In C. P. Nemeth, E. Hollnagel, & S. W. A. Dekker (Eds.), *Resilience engineering perspectives, Volume 2: Preparation and restoration* (pp. 117–133). Ashgate.
Hollnagel, E. (2015). *Introduction to the Resilience Analysis Grid (RAG)* (Technical note).
Hollnagel, E. (Ed.). (2017). Resilience engineering in practice: A guidebook (eBook ed.; 1st ed. 2011). CRC Press. https://doi.org/10.1201/9781317065265
Hollnagel, E. (2018). *Safety-II in practice: Developing the resilience potentials.* Routledge. https://doi.org/10.4324/9781315201023
Hollnagel, E., & Hill, R. (2020). *FRAM Model Visualiser (FMV) instructions* (Version 2.1.4) [Instruction manual]. Functional Resonance Analysis Method.
Hollnagel, E., Woods, D. D., & Leveson, N. (Eds.). (2006). *Resilience engineering: Concepts and precepts.* Ashgate/CRC Press. https://doi.org/10.1201/9781315605685
International Civil Aviation Organization. (2005). Global Air Traffic Management Operational Concept (Doc 9854). https://store.icao.int/en/global-air-traffic-management-operational-concept-doc-9854
Johnson, M. (2020). UTM TIM-04: Conflict management [Presentation]. NASA Technical Reports Server. https://ntrs.nasa.gov/api/citations/20210010271/downloads/UTM%20TIM%2004%20Conflict%20Management%20Johnson%20Final%20MG_redacted.pptx
Johnson, M., & Larrow, J. (2020). UAS Traffic Management Conflict Management Model (Version 2.0.1). NASA Technical Reports Server. https://ntrs.nasa.gov/citations/20205002076
Jung, J., & Craven, N. (2020). Results and analysis from NASA's UTM Technical Capability Level 4 demonstration (NASA/TM-2020-5007032). NASA. https://www.nasa.gov/wp-content/uploads/2024/04/2020-jung-nasa-tm-5007032-508-0.pdf
Lee, J. D., & See, K. A. (2004). Trust in automation: Designing for appropriate reliance. *Human Factors, 46*(1), 50–80. https://doi.org/10.1518/hfes.46.1.50.30392
NASA Aeronautics Research Mission Directorate (ARMD). (n.d.). UTM—NASA technical documents, papers, and presentations [Web page]. https://www.nasa.gov/directorates/armd/aosp/utm/utm-nasa-technical-documents-papers-and-presentations/
NASA Aeronautics Research Mission Directorate (ARMD). (2020, December 10). UTM Overview (v1). NASA Technical Reports Server. https://ntrs.nasa.gov/api/citations/20205011107/downloads/UTM%20Overview_20201210%2C%20v1.pdf
Nemeth, C., & Holbrook, J. (2021). Resilience engineering's potential for advanced air mobility (AAM). In *Proceedings of the 21st international symposium on aviation psychology* (pp. 128–133). Wright State University. https://corescholar.libraries.wright.edu/isap_2021/22/
Parasuraman, R., Sheridan, T. B., & Wickens, C. D. (2000). A model for types and levels of human interaction with automation. *IEEE Transactions on Systems, Man, and Cybernetics—Part A, 30*(3), 286–297. https://doi.org/10.1109/3468.844354
Prinzel, L. J., & Ellis, S. R. (2021, January). Enabling autonomous Advanced Air Mobility: Human-autonomy teaming and in-time system-wide safety assurance. AIAA SciTech 2021. https://ntrs.nasa.gov/citations/20205011262
Reason, J. (1990). *Human error.* Cambridge University Press.
Reason, J. (1997). *Managing the risks of organizational accidents.* Ashgate.
Reason, J. (2000). Human error: Models and management. *BMJ, 320*(7237), 768–770. https://doi.org/10.1136/bmj.320.7237.768
Rios, J. L., Homola, J., Craven, N., Verma, P., & Baskaran, V. (2020). Strategic Deconfliction Performance: Results and analysis from the NASA UTM TCL4 demonstration

(NASA/TM-2020-5006337). NASA. https://ntrs.nasa.gov/api/citations/20205006337/downloads/6337-Strategic%20Deconfliction%20Performance%20in%20TCL4%20v20200909%5B3%5D.pdf

Transport Safety Investigation Bureau. (2024). *Final report: Unmanned aircraft, DrN-35LS, loss of aircraft at sea, 19 February 2024* (TIB/AAI/CAS.230). Ministry of Transport, Singapore.

UK Civil Aviation Authority. (2024). CAP 722H: Unmanned Aircraft Systems — Specific Category Operations: Pre-defined risk assessment requirements, guidance & policy (3rd ed.). Retrieved November 1, 2025, from https://www.caa.co.uk/our-work/publications/documents/content/cap-722h/

University of Utah Kahlert School of Computing. (2018). UTM ConOps v1.0 (mirror of FAA–NASA 2018 document) [PDF]. https://users.cs.utah.edu/~tch/notes/refs/2018-UTM-ConOps-v1.0.pdf

Weick, K. E., & Sutcliffe, K. M. (2007). *Managing the unexpected: Resilient performance in an age of uncertainty* (2nd ed.). Jossey-Bass.

Wickens, C. D., Lee, J. D., Liu, Y., & Gordon-Becker, S. E. (2004). An introduction to human factors engineering (2nd ed.). Pearson Prentice Hall. https://books.google.com/books/about/An_Introduction_to_Human_Factors_Enginee.html?id=nGUItAEACAAJ

5 Monitor
Efficient Performance

INTRODUCTION

Imagine an electric air taxi gliding above a busy city at dawn. Inside the cockpit, the pilot glances at a display showing that all systems are "green," while on the ground, an operations center tracks the flight's every move. Suddenly, a subtle warning light flickers: the battery temperature on one rotor is rising faster than expected. At the same moment, weather sensors indicate a sudden shift in wind patterns ahead. Before any passenger feels a thing, the pilot and the automated systems have already taken action, slowing the aircraft and rerouting slightly to avoid the turbulence. This quick, coordinated response is no accident; it's the result of continuous monitoring that catches issues early and keeps the operation safe. In advanced air mobility (AAM), continuous monitoring is the silent guardian of safety and resilience (see Figure 5.1). In resilience engineering, monitoring essentially means *knowing what to look for*—staying alert to anything that could soon threaten the system's performance. This includes watching the aircraft's internal health (such as battery temperature or other system diagnostics) as well as the external environment, such as weather changes or air traffic conditions. AAM operations involve novel electric vertical takeoff and landing (eVTOL) aircraft, new airspace integration concepts, and complex human–machine teams. In this dynamic context, effective monitoring functions as an early warning system, enabling pilots, controllers, and automated tools to anticipate problems and respond before they escalate into incidents. To achieve resilient and sustainable AAM operations, both technical system health monitoring and human performance monitoring must be addressed.

THE MONITOR FUNCTION IN RESILIENCE ENGINEERING AND FRAM

Resilience engineering considers safety as an emergent property of how systems perform under varying conditions, beyond simply avoiding failures. Erik Hollnagel, a leading scholar in resilience engineering, defines four essential system capabilities (or potentials) for resilient performance: *anticipate*, *monitor*, *respond*, and *learn*. The potential to monitor is the capability of an organization or system to maintain awareness of what is happening that may affect operations in the short term. In Hollnagel's words, it is "knowing what to look for," i.e., being able to monitor those things that could seriously affect the system's performance in the near term, whether positively or negatively. Crucially, this monitoring must encompass both the system's performance and the events occurring in the environment. In practice, *monitoring* means being vigilant to detect irregular

DOI: 10.1201/9781003546306-7

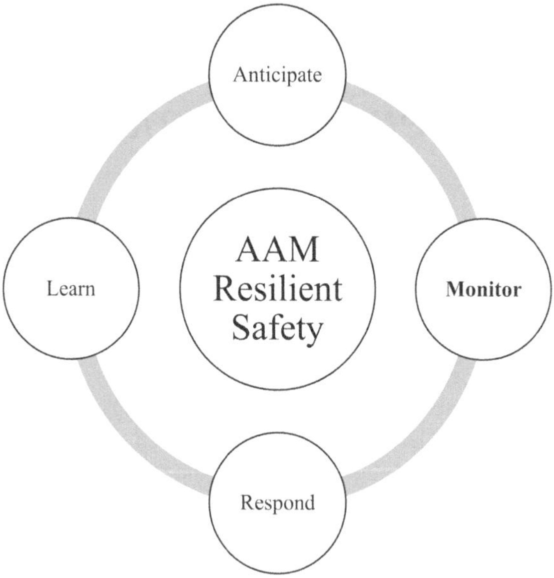

FIGURE 5.1 Monitor function for resilient safety in AAM.

Note: Based on Hollnagel's (2009, 2015) four resilience abilities (anticipate, monitor, respond, learn).

changes, disturbances, or opportunities that could have safety implications, whether they are technical, e.g., a degrading sensor, or contextual, e.g., a sudden weather change. Monitoring complements the other *resilience potentials*; for example, the ability to *respond* to an anomaly relies on first detecting it, and the ability to *anticipate* future risks is informed by trends observed through monitoring.

In FRAM (functional resonance analysis method), monitoring can be modeled as a system function that interacts with others. FRAM is a method to map complex socio-technical systems by defining functions and their interdependencies (inputs, outputs, preconditions, resources, etc.). Monitoring often appears in FRAM models as a function that provides feedback or feedforward information to other functions, helping to dampen variability or amplify adaptations in a controlled way. Hollnagel illustrates the four resilience potentials as interconnected FRAM functions—*monitor*, *respond*, *learn*, and *anticipate*—that influence each other.

In resilience terms, an AAM system with a strong Monitor function is one that knows what to look for and can detect the weak signals of emerging problems. For instance, an operator of an urban air taxi network must continuously monitor each vehicle's health parameters (battery capacity and temperatures, motor and rotor performance, navigation accuracy, etc.) as well as external factors (weather cells developing, GPS signal quality, nearby air traffic, and obstacles) that could affect flight safety. The resilience assessment grid provides a set of questions to evaluate how well an organization supports monitoring, for example: "*Does the organization have real-time visibility into the technical systems' states? Are there defined indicators for when operations may be drifting toward unsafe boundaries? Do operators receive timely alerts of anomalies?*" Such questions help assess the monitor's capability. A key point is that monitoring in complex systems is not merely passive observation; it involves active sensing and sensemaking, requiring the system to filter noise, integrate data, and identify what is genuinely significant for safety.

The Monitor function also plays a prominent role in safety management frameworks. In Leveson's STAMP (systems-theoretic accident model and processes) model, safety is maintained through a hierarchy of control loops, each of which requires feedback to function effectively. STAMP posits that accidents often occur not just from component failures, but from inadequate control or lack of constraint enforcement in the system (for example, a controller not receiving or not correctly processing feedback) (Leveson, 2012). To enforce safety constraints, controllers—whether human or automated—require accurate and timely information about the system state. In other words, effective monitoring and feedback loops are fundamental for each control function. A fundamental STAMP design principle is the need for appropriate feedback in every control structure to keep the controller's mental or internal model of the process up to date. If feedback is missing or delayed, the controller's model can become outdated or incorrect, leading to unsafe decisions. For example, an autonomous eVTOL's flight control computer is a controller that must keep the aircraft within safe limits; it needs continuous sensor feedback (position, speed, battery state, etc.) to do so. At a higher level, an AAM operations center acts as a controller for the fleet, requiring aggregate feedback on each aircraft's status, airspace conditions, and other relevant information. System-theoretic hazard analysis explicitly asks: "*What if the controller does not receive the necessary feedback or receives it too late?*" In AAM, this could translate to questions like: *What if the operator is unaware that a vehicle's sensor has failed? What if an eVTOL's remote pilot does not realize the aircraft is deviating from its assigned route?* Ensuring robust monitoring (and communication) mitigates such scenarios by closing the loop with reliable data.

In summary, resilience engineering and safety science frameworks both identify monitoring as a cornerstone of safe performance. For AAM, this highlights that deploying new aircraft and automation is not enough—there is a need for designing effective monitoring functions into both technology and operations. In the following sections, we examine how this concept is translated into practice, from the sensors and diagnostics that monitor vehicle health to the tools that track the state of the human operator, to the organizational arrangements for continuous safety assurance. The goal is to maintain situational awareness across the system *at all times*, enabling proactive risk management in a dynamic AAM environment.

TECHNICAL SYSTEM HEALTH MONITORING IN AAM

Dallas, 05:42 a.m.—the day's first air-taxi rotation is still 20 minutes away when a maintenance alert flashes on VoloIQ, Volocopter's fleet management portal. Overnight data from the aircraft's digital twin shows that battery cell 42 in pack #3 registered an 8% jump in internal resistance during the last charge cycle, below the flight limit, yet outside the fleet's statistical norm. A cloud-based algorithm, developed under a joint NASA-AFWERX project, spots the outlier within the first 5 minutes of post-charge telemetry, long before a human would notice the drift. The system automatically triggers a *condition-based maintenance* work-order; a technician swaps the module, and the aircraft goes back on the line with zero schedule impact. Passengers never realized their commuter hop was almost launched with an incipient battery fault—proof that continuous health monitoring is the quiet hero of resilient AAM.

Even though this scenario is fictional, it could be a scenario faced by an AAM operator in the near future. As AAM aircraft depend on tightly integrated avionics, electric or hybrid-electric propulsion, high-energy batteries, and software-centric flight controls, and to keep this complexity reliable, operators layer multiple monitoring technologies:

These layers transform system monitoring from a passive log-book exercise into an active, predictive safety net. Field data already confirm the payoff: battery-stress studies show that rapid-discharge eVTOL cycles shorten cell life unless real-time thermal and electrical data are used to manage charge windows; NASA-funded weak cell detection now flags problematic modules 25 minutes sooner than legacy checks.

Looking forward, 5G-enabled edge architectures promise millisecond latency links between aircraft and local compute nodes, meeting the stringent networking requirements identified for UAM traffic. As fleets grow, such connectivity will allow distributed digital twins to run short-horizon *what-if* simulations for every flight, pushing alerts to operators before a fault propagates. Recent research even explores AI-driven, low-altitude networks that adapt routing and compute resources dynamically to maintain those monitoring loops under heavy load.

In short, technical health monitoring in AAM is no longer about capturing data for the post-flight archive; it is about *seeing the future early enough to change it.* Whether it is a battery module swapped at dawn in Dallas, an EHang route automatically re-planned in Guangzhou, or a Joby test article generating terabytes of vibration data in California, the common thread is the same: watch everything, understand what it means, and act before passengers—or the public—ever feel a ripple.

Vehicle Diagnostic Systems and On-Board Monitoring

Modern aircraft include built-in test equipment and health monitoring units, and AAM vehicles are no exception. The eVTOL designs typically incorporate sensors throughout the aircraft to track the status of critical components, such as motors, rotors, battery cells, and flight control actuators. For example, battery management systems continuously monitor battery temperature, voltage, and current to manage power usage and detect any signs of thermal runaway or cell degradation.

Flight control computers monitor attitude and navigation sensors; if anomalies (like oscillations or bias errors) are detected, fail-safe logic can engage to maintain stability. EHang's autonomous aerial vehicles feature a built-in *Fail-Safe* function in the flight control system that "evaluates the health condition of the aircraft in real-time." This system monitors each onboard device and subsystem and transmits the information back to the command-and-control center. In case of any abnormal condition, the system will automatically select and execute the smoothest route based on preset logic "to safely take the passenger to the destination." In other words, the aircraft itself performs first-line monitoring and can initiate contingency actions (like an emergency landing or reroute) upon detecting a critical failure. This level of autonomy in monitoring is crucial, given that many AAM aircraft are designed to operate with minimal or no onboard pilot intervention.

Real-Time Telemetry and Ground Monitoring

Robust data links to ground control centers complement on-board monitoring. A defining feature of AAM operations is the extensive use of telemetry—continuous streaming of flight data to operators on the ground. Through high-bandwidth wireless links (e.g., LTE/5G networks or satellite), parameters such as position, altitude, speed, attitude, system temperatures, battery charge, motor RPM, flight mode status, and other relevant data are transmitted to a control center in real-time. This enables ground operators and automated systems to have a real-time view of each aircraft's health and progress. If an issue is developing, ground control can alert the flight crew (if manned) or send remote commands (if unmanned) to investigate or mitigate it. For instance, EHang's passenger-grade AAVs utilize 4G/5G wireless links to enable "smooth communication" between the UAV and the command center, achieving remote control of the aircraft and real-time transmission of flight data.

Many AAM companies have developed centralized operations centers akin to a mission control. These centers display the real-time telemetry of multiple vehicles on screens and dashboards, often overlaying aircraft positions on maps and showing key indicators (engine or battery status, warning messages, etc.). They function as the nerve center for fleet monitoring, dispatch, and emergency management. For example, Joby Aviation reportedly monitors all test flights via a dedicated control room with engineers watching telemetry and voice comms; Volocopter in Germany has its *VoloIQ* digital platform (discussed later), which will be used by staff to oversee flights; and China's EHang has a highly integrated command-and-control center for its autonomous aerial vehicles. These facilities are essential, especially for beyond-visual-line-of-sight (BVLOS) operations where vehicles fly out of direct sight; continuous electronic surveillance is the only way to track and manage them.

In traditional aviation, health monitoring data (such as engine performance trends or parameter exceedances) may be downloaded post-flight for analysis (e.g., through FOQA—flight operational quality assurance programs). In AAM, however, the model is shifting toward real-time health management and intervention. Operators aim to identify and address issues promptly. For instance, when EHang began initial commercial operations, the company agreed with regulators to impose prudent limits and *closely monitor new routes during the first phase of operations*, even positioning extra observers along out-of-sight segments, to ensure any unexpected issues are immediately caught and managed. This practice is partly to satisfy regulators for early deployments, but it also sets the stage for long-term, continuous monitoring of operations at scale. By adopting a highly active monitoring posture, AAM operators and authorities can gather data and establish trust in the system before scaling up the frequency and autonomy of flights.

Digital Twins and Predictive Maintenance

A significant innovation in system monitoring is the use of digital twins—virtual replicas of physical assets that run in parallel with the real system, receiving real-time data. In AAM, a digital twin of an eVTOL can simulate its behavior under current conditions and even project future states, allowing predictive analytics.

Digital twins have been identified as a key enabling technology for AAM certification and operations. One recent study notes that "*Digital twins can be used to monitor the health and performance of AAM systems, predicting maintenance needs and reducing downtime.*" By comparing sensor data from the actual vehicle with expected parameters in the digital model, anomalies can be spotted early. For example, suppose the digital twin of a motor predicts a specific vibration pattern, but the real motor starts deviating from that pattern. In that case, maintenance can be scheduled before a failure occurs. This is the essence of *predictive maintenance*. Using AI and machine learning algorithms to analyze trends from large datasets, operators can identify leading indicators of component wear or risk (for instance, a subtle increase in battery internal resistance or a pattern of short voltage drops under load) and act on them proactively. In the highly distributed AAM environment, predictive maintenance will be crucial—vehicles will operate in networks where pulling one out of service unexpectedly could disrupt schedules. Early fault detection via smart monitoring minimizes such surprises. NASA and industry have emphasized this: for example, Airbus reports that its predictive maintenance programs have reduced unscheduled removal events by 10–50% for various aircraft systems. We can expect similar or greater benefits for eVTOL fleets, which will rely on electric propulsion and fast turnarounds. By leveraging big data from many flights, AAM operators can not only react to failures but also anticipate them. In fact, global data-sharing initiatives are emerging: Aireon's space-based ADS-B surveillance network, for instance, introduced a *Safety Dashboard* tool to help users identify, monitor, and analyze key leading indicators of risk in their airspace, with a visual display that makes it easy to spot trends, patterns, and outliers. Such capabilities enable a shift from reactive to predictive safety management.

Digital twins can also model the operational environment, not just the vehicle. For instance, an airspace or city digital twin might ingest data about weather, communications networks, and traffic flows, effectively monitoring the context for each flight. Volocopter's VoloIQ platform is described as providing *airspace digital twins* and vehicle data logging and analysis as part of its standard UAM operating system. In practice, this means VoloIQ will continuously mirror each flight in a virtual model that tracks the vehicle's condition and the airspace situation, allowing the operator to foresee conflicts or constraints. The aim is to provide "*complete digital transparency and greater ecosystem efficiency in real time.*" By integrating these elements, AAM operators can maintain a comprehensive, real-time digital picture of operations.

Edge Computing and Sensor Fusion

AAM aircraft generate massive data streams from sensors such as LiDAR, cameras, radar, inertial measurement units, and more, especially autonomous models that require detect-and-avoid capabilities. Transmitting all raw data to the ground is impractical due to bandwidth and latency constraints. Edge computing refers to processing data on board (or at a nearby edge node) to distill high-rate sensor data into meaningful events or indicators. For monitoring, edge computing enables faster anomaly detection and response without waiting for ground analysis. For example, an onboard edge-AI algorithm might monitor a camera feed to detect obstacles (such as birds or other drones) and only send an alert or processed observation to ground control, rather than streaming the entire video. This reduces information overload and focuses attention on significant signals. Similarly, sensor fusion combines multiple sensor inputs to improve the reliability of monitoring. An eVTOL's avionics might fuse GPS and inertial measurements to detect if one source is errant, or combine motor current, temperature, and vibration readings to diagnose a developing propulsion issue. By fusing sensors, the system can monitor higher-level states that single sensors alone cannot reveal (such as overall *propulsion health* or *flight stability*). Emerging AAM designs are exploring redundant sensor architectures whose outputs are continuously cross-checked; when a disagreement or out-of-bound condition occurs, the system flags it for investigation or triggers a contingency.

In summary, technical monitoring in AAM involves a multi-layered approach: on-board systems watch component health and flight parameters; robust communication links send data to ground stations; control centers aggregate and visualize the data for human supervisors; and advanced analytics (AI/digital twins) process the data to find subtle trends and predict future issues. All these layers work together to ensure that "nothing falls through the cracks." Ideally, a failure or deviation in an AAM vehicle will be detected either by the vehicle's diagnostics or by the remote operations team well before it becomes a safety-critical issue. This proactive stance—continuous monitoring and early intervention—is what enables resilience in complex operations. A key challenge, however, is avoiding information overload, as operators may be overwhelmed with data from dozens of vehicles. Thus, intelligent filtering, alerting, and user interface design become crucial, as discussed in the next section regarding the human factors of monitoring. The necessary technological capabilities are in place, including real-time data streaming, big data analytics, and AI-driven anomaly detection. The task is to integrate them in a way that ensures the *right* information reaches the *right* people or systems at the *right* time—the essence of effective monitoring.

HUMAN PERFORMANCE MONITORING IN AAM OPERATIONS

In any aviation system, humans play integral roles—as pilots, remote operators, air traffic managers, maintenance technicians, and more. In AAM, humans will interact with new automation and face novel operational contexts (e.g., managing many autonomous aircraft simultaneously from a control station). Ensuring that human performance remains optimal is a critical part of the Monitor function. This section examines how we monitor the *human* element in AAM: from assessing pilot workload and cognitive state to tracking the performance of UAM air traffic management personnel and remote supervisors. It also considers *human-machine teaming*—i.e., monitoring how effectively humans and automation work together. By monitoring human performance, we can promptly detect signs of overload, fatigue, confusion, or error, allowing us to adjust procedures or provide support and thereby prevent incidents attributable to human factors.

Pilot Workload and Cognitive State Monitoring

Early urban air mobility operations will often involve a human pilot (onboard or remote) to ensure safety while automation matures. These pilots may face high workloads due to managing new vehicle types and dense low-altitude airspace with novel procedures. Monitoring pilot workload and cognitive readiness is, therefore, crucial. Traditional measures of pilot workload include subjective ratings (such as the NASA-TLX or Bedford workload scales) and performance metrics (response times, error rates); however, physiological monitoring is also being increasingly used. Researchers have begun equipping pilots with wearable and cockpit-integrated sensors, such as eye-tracking glasses, EEG (electroencephalogram) brainwave monitors, and heart rate sensors, during simulated or actual AAM flight scenarios to gauge workload and attention in real-time (Masi et al., 2023). For example, during NASA's 2024 *Integration of Automated Systems* flight test campaign, pilots flew high-density UAM scenarios while researchers continuously gathered biometric data—eye-tracking to pinpoint visual focus, mobile EEG to measure brain activity, and heart-rate sensors to assess physiological stress—yielding rich insights into workload and cognitive engagement under advanced automation (Monk et al., 2024; Mitchell, 2023). Such data can reveal if a pilot is overloaded (e.g., erratic eye scan patterns, spikes in heart rate, or EEG indicators of task saturation) or conversely underloaded and disengaged (which can lead to loss of situational awareness).

Real-time pilot state monitoring could be used operationally. For instance, a cockpit-mounted camera tracking a pilot's gaze might detect that the pilot has not checked an important instrument or display for an extended period and trigger an alert to refocus their attention. Similarly, if a wearable EEG or eye-blink monitor detected early signs of fatigue or microsleep, the system could issue a caution or suggest a break (applicable for remote operators or future single-pilot scenarios). While these technologies are still maturing, they represent a proactive approach to managing human performance. Rather than waiting for a pilot error to manifest due to overload or fatigue, the system monitors the precursors (workload indicators) and provides support or mitigation. This ties into the resilience principles of *anticipating* and *responding* to issues before they fully manifest.

Research in this area underlines the value of combining multiple measures. In one recent experiment, pilots flew challenging AAM scenarios while both objective and subjective workload data were collected. The researchers found that combining physiological metrics with self-assessments provided a more comprehensive picture of the pilot's state (Monk et al., 2024). Advanced sensor fusion for human monitoring is on the horizon. For example, one concept is to integrate voice analysis, eye tracking, and heart rate variability to continuously assess a pilot's cognitive load and stress. If such a system detected a pattern consistent with overload, it could proactively simplify the pilot's tasking or engage an autopilot mode to assist. Conversely, in low workload situations, it might periodically "check in" with the pilot to ensure vigilance. Interestingly expanding the possibilities in the field, a recent study presented a low-interference workload-recognition framework that fuses voice (speech) features with biometric sensors—eye-tracking and ECG (electrocardiogram)—to analyze pilot communications and scan patterns, detect cognitive strain, and trigger machine-based support when overload is imminent (Xu et al., 2025).

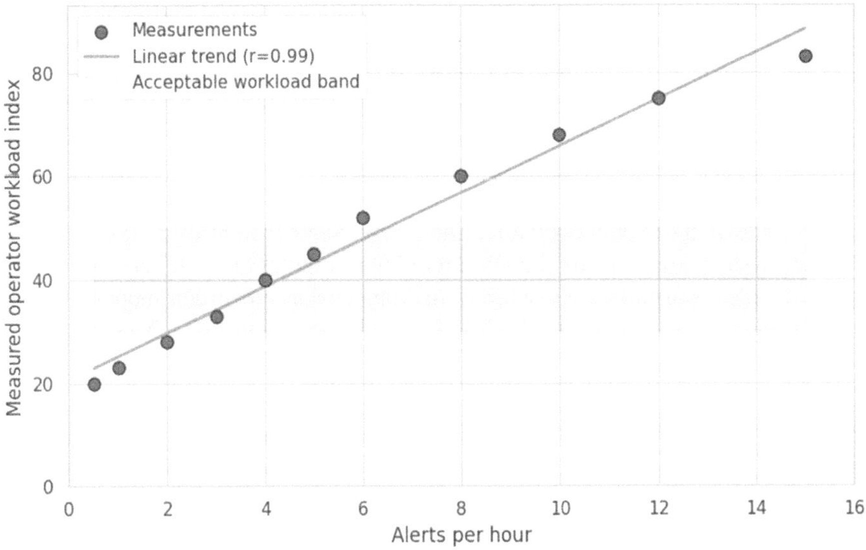

FIGURE 5.2 Operator workload vs. alert density.

An example of *operator workload index* vs. *alerts per hour* is depicted in Figure 5.2, which could be helpful to identify how *alert frequency* impacts *operator cognitive load* so teams can set alert-rate thresholds, prioritize or suppress alerts, and design staffing/training to keep workload within the acceptable band.

Monitoring of Remote Operators and UAM Traffic Managers

Many AAM concepts involve ground-based operators who supervise flights, particularly for autonomous air taxis or cargo drones. These could be remote pilots who intervene only as needed, or fleet managers overseeing multiple semi-autonomous vehicles. Additionally, in high-density UAM corridors, air traffic management may rely on automated services, but human controllers or supervisors will still be involved to handle off-nominal situations. Monitoring the performance and workload of these personnel is as important as it is for traditional pilots. A remote pilot station can log every control input, the timing of responses to alerts, and other interaction metrics. Suppose a single remote operator is responsible for several autonomous drones. In that case, the system might monitor how frequently they have to intervene or how long it takes to acknowledge alerts. If interventions per hour or response times start to spike, it could indicate an underlying issue (perhaps the autonomy is struggling with conditions, leading to operator overload).

Likewise, for air-traffic controllers or U-space managers who must juggle mixed conventional and UAM traffic, cognitive load can be gauged through proxy indicators such as the rate of radio transmissions and the number of active conflicts, supplemented—when needed—by physiological measures (Manning et al., 2002; Muñoz-de-Escalona et al., 2025; Yang et al., 2023). During the SESAR-funded CORUS-XUAM very-large demonstrations, the U-space service provider's conformance-monitoring tool tracked every UAM trajectory in real time and automatically compared it with the corresponding flight plan; any deviation or airspace-constraint breach triggered an alert to the operator and a notification to the relevant authority (CORUS-XUAM Consortium, 2022; SESAR Joint Undertaking, n.d.; SESAR JU, 2022). This automation relieves human controllers from manually inspecting each flight yet still leaves them responsible for evaluating and acting on the alert, underscoring the need for alert-filter logic that supports rather than overwhelms situational awareness (Guillermet, 2019; SESAR Joint Undertaking, 2022). Current system designers are therefore refining thresholds and rule sets so that only safety-relevant warnings are displayed, striking a balance between supportive automation and information overload (SESAR Joint Undertaking, 2023; Xu et al., 2025).

United States: NASA and FAA—Data-Driven Monitoring and In-Time Safety Assurance

In the United States, NASA and the FAA are building complementary frameworks that move safety assurance into real time, using dense telemetry, predictive analytics, and cross-agency data sharing. NASA's System-Wide Safety program pioneered the *In-Time System-Wide Safety Assurance* concept, which envisions algorithms that monitor every flight, detect precursors to risk, and recommend mitigations before an incident occurs (Ellis & Koelling, 2020). A prototype *in-time safety monitoring and prediction* tool—initially tested on low-altitude drones—ingests live telemetry, spots trends such as narrowing separation or drifting system parameters, and pushes alerts to operators (Corbetta et al., 2019).

The FAA is reinforcing that vision through long-standing data-sharing programs such as ASIAS, which mines voluntarily submitted flight data to identify hazards before they manifest (U.S. Department of Transportation, Office of Inspector General, 2021.). The Agency's Innovate28 implementation plan commits to the same data-driven approach that created one of the aviation systems in the world, emphasizes phased entry, continuous monitoring and iterative learning during early AAM deployments (FAA, 2023a).

A concrete example of institutional monitoring is the FAA–U.S. Air Force data-sharing agreement: in October 2023, AFWERX Prime and the FAA signed an MoU to exchange eVTOL flight-test data, accelerating standards development and exposing failure modes early (Tresslar, 2023). Military agility prime sorties thus flow directly into the FAA's safety databases, closing a critical feedback loop.

NASA's AAM National Campaign extends in-time monitoring to full-scale flight trials with Joby, Wisk, and others. Every aircraft carries high-rate instrumentation, and NASA links those streams to a research *network ops center* that records anomalies, automation modes, and pilot inputs—providing a living laboratory for detect-and-avoid, noise, and contingency procedures (Bagnall & West, 2023). Early simulations revealed that *automation surprises* could catch pilots off guard; consequently, researchers now recommend that future AAM interfaces display the automation's intended actions alongside the aircraft's state, allowing operators to monitor both (National Aeronautics and Space Administration, 2022).

Parallel work at the U.S. DOT Volpe Center explores how a metropolitan operations center might integrate these data flows. In 2022–2023, Volpe built a notional *UAM Dashboard* that fuses vehicle telemetry, weather feeds, surveillance tracks, and demand-management data, demonstrating that a single interface can monitor an entire city's worth of AAM flights (U.S. DOT Volpe Center, 2024). FAA officials emphasize that early AAM operations will initially take place in restricted corridors, specifically to collect baseline data before scaling up (U.S. DOT Volpe Center, 2024).

To institutionalize real-time data flow, the FAA is deploying the flight information management system (FIMS). This interface enables UAM operators to exchange flight plans, position, and status data with the National Airspace System. FIMS is defined in the FAA's UTM ConOps v2.0 as the FAA-managed interface supporting data exchange between UTM participants and FAA systems (including post-hoc querying and contingency notifications). In the UAM ConOps v2.0, the analogous FAA–industry gateway is described as the FAA-Industry Data Exchange Protocol, through which FAA systems can request UAM operational data on demand and distribute FAA information to UAM stakeholders (FAA, 2020; FAA, 2023b).In practice, this shifts monitoring from radar-only surveillance to a cooperative, network-centric model where every AAM flight continuously shares its state and intentions.

Together, NASA's predictive analytics research, the FAA's data-sharing infrastructure, Air Force flight-test contributions, and Volpe's operational simulations form a layered U.S. ecosystem for in-time safety assurance—a real-time dynamic *risk meter* for AAM.

Human-Machine Teaming and Trust Monitoring

AAM operations will rely heavily on human–machine teaming—whether a pilot collaborating with an adaptive autopilot, a remote supervisor orchestrating several autonomous drones, or a maintenance engineer drawing on AI-driven diagnostic advice. Monitoring the *team* (human + automation) is therefore emerging as a critical discipline. It requires assessing how well the automation supports the human *and* how effectively the human supervises or intervenes with that automation (Chancey et al., 2021).

A growing body of research shows that trust and workload-sharing are pivotal metrics. Excessive trust can push an operator "*out of the loop*," while mistrust can prompt needless double-checking and mode reversions (Parasuraman & Riley, 1997). To quantify these states, investigators have trialed *real-time trust sensors* that fuse speech prosody, eye-tracking, and physiological measures to infer an operator's trust level and let the automation adapt accordingly (Hu et al., 2016).

Established human-factors tools, such as crew resource management (CRM), are being adapted for distributed AAM crews. The FAA's *Remote Pilot Study Guide* already recommends CRM checklists—even when the "crew" is a single remote pilot plus a visual observer linked by data and voice (FAA, 2021). Researchers have extended these ideas to large UAS and multi-aircraft control, outlining briefing, call-out, and cross-monitoring techniques tailored to remote operations (Neff, 2019).

Automation can support CRM tasks, too. NASA cockpit-aid prototypes monitor pilot switch selections and issue caution-and-warning alerting for abnormal situations

(Hanaway & Moorehead, 1989). Similar concepts are migrating to the AAM control-center software. In case the system detects that a supervisor has not acted on a high-priority alert—perhaps because they are managing several other flights—it can escalate the warning to another controller or invoke a safe-state command.

Physiological monitoring is also beginning to extend from flight decks to operations rooms. Recent work on remote-tower air traffic controllers combined EEG, eye tracking, ECG, and electrodermal activity to model visual fatigue and trigger fatigue recognition and monitoring when thresholds were exceeded (Liang et al., 2025). Agencies such as NASA are even piloting quick cognitive-readiness checks at shift start to verify that AAM operators are fit for duty (Monk et al., 2024), although privacy and acceptability remain open questions.

All these monitoring efforts serve one overarching goal: to keep human performance within safe limits and spot drift early. History shows that lapses in attention, overload, or miscommunication frequently precipitate incidents (Histon & Wickens, 2010). By embedding smart monitoring, whether through adaptive software "wingmen," CRM-style cross-checks, or unobtrusive biometrics, AAM enterprises can intervene before a slip becomes an accident. Moreover, a Safety-II outlook reminds us that we should also study the occasions when human–machine teams perform brilliantly under stress; that positive data feeds the *Learn* function of resilience just as surely as anomaly reports (Hollnagel, 2014).

AAM MISSION SAFETY INDEX (AMSI)

AMSI sounds *ámesi* (*άμεση*) — "*immediate*" in Greek, showcasing the need for rapid response to any safety-related issue that arises during a mission within the AAM ecosystem. AMSI is a single, operationally usable score (0–100) to support go/no-go and continual in-flight risk management for AAM air transport missions by aggregating the five 5M pillars—man, machine, medium, mission, and management. Built on FAA (2022, 2024) flight risk assessment tool (FRAT) principles, e.g. risk identified and mitigated before and during flight, and designed for live data feeds.

It is computed as "*100 minus the sum of weighted risk points*," where each pillar carries an agreed weight (typically equal by default and adjustable by the operator). As depicted in Figure 5.3, scores at or above 80 are *Green* and may proceed under normal authority; scores from 60 to 79 are *Amber* and require an operations supervisor or duty pilot to review and record mitigations; scores below 60 are *Red* and the mission does not launch unless a safety manager authorizes specific mitigations that lift the score back to an acceptable level.

Before flight, AMSI-Pre is determined by examining each pillar and assigning a score from 0 to 10 to the highest applicable condition within that pillar, then multiplying by the pillar's weight. For example, as illustrated in Figure 5.4, *Man* covers

AAM MSI Band & Actions	
Green (≥80)	Proceed within normal authority.
Amber (60–79)	Duty pilot/ops supervisor review; document mitigations.
Red (<60)	No-go unless safety manager authorizes with specific mitigations.

FIGURE 5.3 AMSI bands and actions.

Pillar (5M)	Factor	Contribution Score (0: Low - 10: High)	Mitigation / Notes
Man	Currency & recent experience (type/route/task)	7	
Man	Duty/Rest & fatigue self-assessment	1	
Man	Human-automation interaction history (last 14 d)	7	
Machine	Deferred items / MEL-equivalent	1	
Machine	Critical faults last 30 d (propulsion/energy/controls/sensors)	0	
Machine	Energy SoH & pre-dep test margins	0	
Machine	Automation & sensing health (radar/vision/RTK)	1	
Medium	Weather vs minima (wind/gust/ceiling/vis)	3	
Medium	Terrain/obstacles & launch/landing site condition	3	
Medium	Airspace context (GEO/NOTAM/TFR)	1	
Mission	Payload & CG vs limits (incl. chemical/load profile)	1	
Mission	Task complexity & proximity (people/structures/roads)	4	
Mission	Route/contingencies & alternates	3	
Mission	Endurance profile vs plan (reserve ≥20% SoC)	2	
Management	Authorizations/permissions (waivers, GEO unlocks, landowner)	1	
Management	SOP/JHA completed & briefed	1	
Management	Emergency readiness (spill kit, extinguisher, egress)	1	

FIGURE 5.4 Example of AMSI-Pri contributory factors.

crew currency and fatigue, *Machine* covers deferred items and system health, including energy and automation, *Medium* covers weather, terrain, and airspace context, *Mission* covers payload, route, alternates, and endurance against the reserve policy, and *Management* covers permits, SOP readiness, and emergency preparedness. Any selection of 5 or 10 requires a named mitigation, and the final AMSI-Pre, along with approvals, is recorded in the SMS log.

During flight, the score becomes AMSI-Live. It starts from AMSI-Pre and automatically adjusts as telemetry, airspace, weather, and infrastructure events occur. Conflict advisories or loss-of-well-clear reduce the score; automation drop-outs or critical system faults reduce it further; energy reserve shortfalls and deteriorating weather also subtract points (see Figure 5.5). If AMSI-Live falls into Amber (see Figures 5.6 and 5.7)

Category	Trigger	Delta (subtract from AMSI)	Action / SOP
GEO/Airspace	Enhanced-Warning area entry (mission-necessary)	−5	Confirm authority; minimize dwell; alternate if possible
GEO/Airspace	Authorization/Restricted zone warning affects track	−15	Hold/land; obtain unlock/authorization or reroute
Traffic	≥2 proximity/conflict advisories within 10 min	−10	Exit hotspot; change altitude/route
Automation	ATTI reversion (loss of GNSS/RTK)	−15	Recover to safe hover/land at nearest suitable
Automation	Obstacle-avoidance brake/hover <10 m or repeated bypass failure	−10	Reduce speed/height; reassess route
Machine	Single-point propulsion/energy channel fault	−15	Nearest suitable landing if persisting
Energy	Predicted landing reserve < 20% SoC	−20	Mandatory divert/alternate
Energy	Reserve trending −5% within 10 min	−5	Optimize route/power; reassess alternate
Weather	Winds/LLWS exceed limits on route	−15	Reroute or land

FIGURE 5.5 Example of live triggers to AMSI.

Metric	Value	Mitigation
Date (UTC)	*2025-11-01 04:58*	
Aircraft	AAM Model 1	
Route / Block	R1/B1	
Alternates	R2/B2/V2	
Man	7	Remote GSO
Machine	1	Nil
Medium	3	Limited access to the TOL site
Mission	4	Remote GSO, Preparation
Management	1	
Weighted points (auto)	36,5	
AAM MSI (Mission Safety Index)	63,5	
AAM MSI Band	AMBER	

FIGURE 5.6 Example of AMSI-Live.

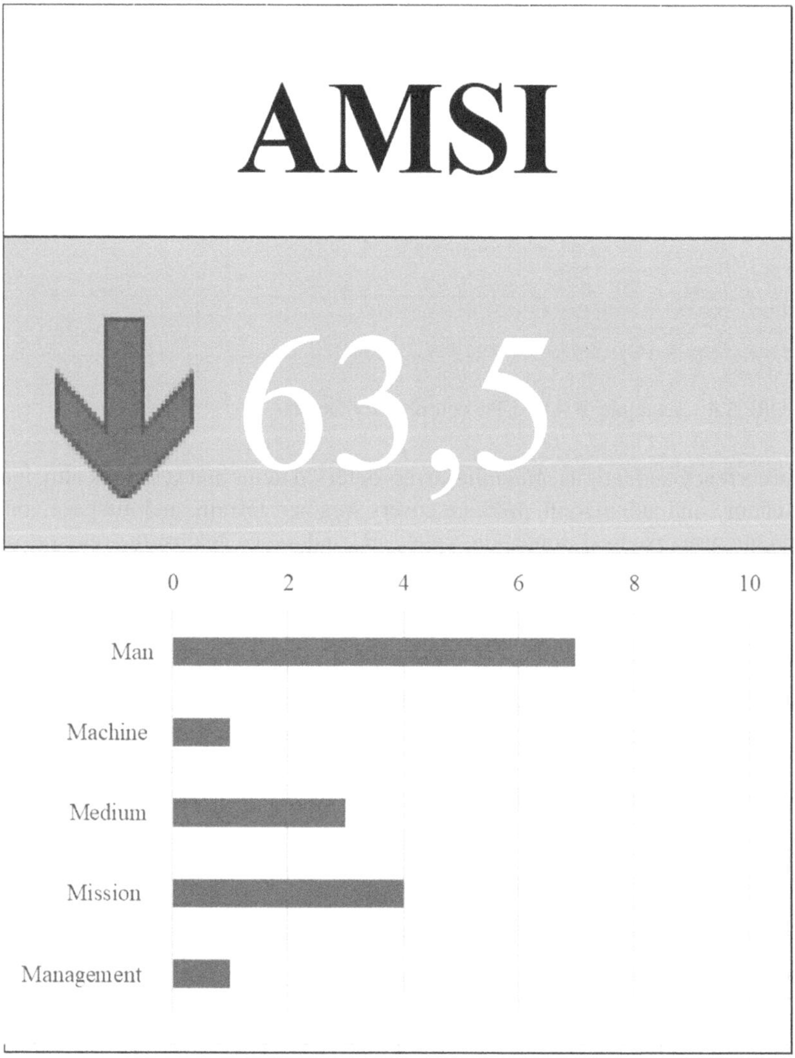

FIGURE 5.7 AMSI dashboard.

or Red, the CCC prompts the crew with the required actions; repeated Amber events trigger coordination with the operations center, while any Red-class combination that breaches hard limits leads to an immediate precautionary landing at the nearest suitable site.

After flight, the AMSI timeline and any mitigations are attached to the journey log for review. The operator can periodically recalibrate pillar weights and thresholds to keep the measure honest as data accumulates, maintaining a conservative stance with dual-source data where practical and ensuring that human decision-making remains in the loop for any automated downgrade.

CASE STUDY: EHANG AND CAAC—CENTRALIZED FLEET MONITORING AND PHASED OPERATIONS

China's advanced air mobility story is inseparable from EHang's autonomous passenger drones and the Civil Aviation Administration of China's (CAAC) close-coupled oversight. Every passenger flight is streamed in real time to EHang's Guangzhou command-and-control center (CCC), whose five-function architecture—monitoring, dispatching, controlling, early warning, and centralized management—is the linchpin of a *progressive-risk* rollout endorsed by CAAC. The regulator's phased approvals, mandatory field observers, and city-scale geo-fenced test zones illustrate how China is using intensive monitoring to scale operations while containing risk, an approach now being extended to other operators, such as JD.com's logistics network and the Shenzhen municipal UTM platform.

China's AAM landscape has been heavily shaped by EHang, the country's flagship autonomous-air-taxi firm (EHang, n.d.-a). EHang's network is orchestrated through an Intelligent Command-and-Control Center (ICCC) that streams telemetry, battery status, and even live video from every aircraft, enabling operators to issue remote reroutes or immediate landing orders if early-warning algorithms flag a problem (EHang, n.d.-b).

In October 2023, CAAC issued the world's first passenger-eVTOL type certificate to the EH216-S, launching commercial service under a *progressive-risk* framework that requires extra observers and route-by-route data collection in the initial phase (EHang, 2023; Aero-News Network, 2023). Company statements underline that "EHang will closely monitor new routes during the first phase of commercial operations," positioning observers along any BVLOS segments to supplement command-center oversight (CompositesWorld, 2023).

Field monitoring is reinforced by geo-fenced *safety zones*. Earlier flight test campaigns in Guangzhou demonstrated that the EH216 automatically detected virtual fences and returned or landed before breaching restricted airspace, earning a CAST safety level II certificate (Vertical Magazine, 2019). At the national scale, CAAC is deploying the UAS Operation Management System (UOMS) to integrate low-altitude tracks with conventional ATC; policy papers describe UOMS as the backbone for registration, alerting, and live-position services (Butterworth-Hayes, 2024).

Cities are implementing their own layers. Shenzhen's municipal UTM now supervises more than 2,000 daily drone flights through a city-wide monitoring platform that shares live data among police, air-traffic managers, and operators (Butterworth-Hayes, 2019; Shenzhen Government Online, 2018). Such hubs align with China's preference for tightly controlled airspace and provide a template for monitoring larger AAM fleets.

EHang's human-performance model is equally centralized: all remote pilots work in one room under shift supervisors, with every action logged for CAAC audit. The regulator also requires periodic performance reports that document anomalies and responses, effectively "monitoring the monitoring" before permitting expansion (EHang, 2023a, b).

Other Chinese ventures echo this pattern. JD.com's heavy-cargo and last-mile drone programs run through a cloud-based control center that integrates flight plans, real-time tracking, and maintenance analytics, mirroring the centralized philosophy that now characterizes China's *low-altitude economy* build-out (JD.com, 2017).

Together, these elements—command center telemetry, phased certification, geo-fenced airspace, and city-wide UTM—illustrate a uniquely Chinese model: scale cautiously, monitor everything, and only loosen restrictions when data proves reliability.

REFERENCES

Aero-News Network. (2023, November 2). *EHang EH216-S eVTOL nears production certification.*

Bagnall, T., & West, L. (2023, September 29). *National Campaign (NC) Tech Talk: Integrated Data Product* [Presentation]. NASA Aeronautics Research Mission Directorate. https://ntrs.nasa.gov/api/citations/20230013169/downloads/AOM%20TechTalk-IDP-FINAL.pdf

Butterworth-Hayes, P. (2019, March 16). *Shenzhen hosts over 2,000 drone flights a day after UTM system launch. Unmanned Airspace.*

Butterworth-Hayes, P. (2024, June 4). *China accelerates its plan to develop a USD44 billion low altitude economy by 2026. Unmanned Airspace.*

Chancey, E. T., Politowicz, M. S., Sato, T., & Yamani, Y. (2021). *Human factors in the emerging market of advanced air mobility* [Prerecorded conference presentation]. *21st International Symposium on Aviation Psychology.* NASA Technical Reports Server.

CompositesWorld. (2023, October 30). *EHang progresses operations, commercial deployment of EH216-S.* https://www.compositesworld.com/news/ehang-progresses-operations-commercial-deployment-of-eh216-s

Corbetta, M., Banerjee, P., Okolo, W. A., Gorospe, G. E., & Luchinsky, D. G. (2019). *Real-time UAV trajectory prediction for safety monitoring in low-altitude airspace* (NASA Technical Memorandum 20190027599). NASA Ames Research Center. https://ntrs.nasa.gov/api/citations/20190027599/downloads/20190027599.pdf

CORUS-XUAM Consortium. (2022, July 13). *U-space Concept of Operations (ConOps), edition 3.10* (Deliverable D4.1). EUROCONTROL / SESAR Joint Undertaking project deliverable.

EHang. (n.d.-a). *EH216-S.* Retrieved January 24, 2026, from https://ehang.com/ehang216s/

EHang. (n.d.-b). *UAM – Passenger Autonomous Aerial Vehicle (AAV).* Retrieved January 24, 2026, from https://www.ehang.com/ehangaav

EHang. (2023a, October 30). *EHang continues to promote operations and commercial deployment of EH216-S passenger-carrying unmanned aerial vehicle system.*

EHang. (2023b, August 21). *EHang unmanned aircraft cloud system approved by the CAAC; EH216-S positioned for post-TC commercial operations.*

Ellis, K., & Koelling, J. (2020, June). *In-time System-wide Safety Assurance (ISSA) concept of operations and design considerations for Urban Air Mobility (UAM)* (NASA/TM—20205003981). NASA Langley Research Center.

Federal Aviation Administration. (2020, March 2). *Unmanned Aircraft System (UAS) Traffic Management (UTM) concept of operations (ConOps), version 2.0.* U.S. Department of Transportation, Office of NextGen.

Federal Aviation Administration. (2021, February 1). *Advisory Circular 107-2A: Small Unmanned Aircraft Systems (sUAS).*

Federal Aviation Administration. (2022, February 4). *Flight Risk Assessment Tools* (Fly Safe fact sheet). https://www.faa.gov/newsroom/safety-briefing/flight-risk-assessment-tools

Federal Aviation Administration. (2023a, July). *Advanced Air Mobility (AAM) implementation plan: Near-term (Innovate28) focus with an eye on the future of AAM* (Version 1.0). https://www.faa.gov/sites/faa.gov/files/AAM-I28-Implementation-Plan.pdf

Federal Aviation Administration. (2023b, April 26). *Urban Air Mobility (UAM) concept of operations, version 2.0.* U.S. Department of Transportation, Office of NextGen.

Federal Aviation Administration. (2024, October 28). *Flight Risk Assessment Tool (FRAT) – FAA Safety Team.* https://www.faa.gov/general/flight-risk-assessment-tool-frat-faa-safety-team

Hanaway, J. F., & Moorehead, R. W. (1989). *Space shuttle avionics system* (NASA SP-504). National Aeronautics and Space Administration.

Hollnagel, E. (2009). The four cornerstones of resilience engineering. In C. P. Nemeth, E. Hollnagel, & S. W. A. Dekker (Eds.), *Resilience engineering perspectives, Volume 2: Preparation and restoration* (pp. 117–133). Ashgate.

Hollnagel, E. (2014). *Safety-I and safety-II: The past and future of safety management.* Ashgate.

Hollnagel, E. (2015). *Introduction to the Resilience Analysis Grid (RAG)* (Technical note).

Hu, W.-L., Akash, K., Jain, N., & Reid, T. (2016). Real-time sensing of trust in human–machine interactions. *IFAC-PapersOnLine, 49*(32), 48–53. https://doi.org/10.1016/j.ifacol.2016.12.188

JD.com. (2017, August 29). *JD.com announces a series of new agreements for drone development.* JD Corporate Blog. https://jdcorporateblog.com/jd-com-announces-series-of-new-agreements-for-drone-development/

Leveson, N. G. (2012). *Engineering a safer world: Systems thinking applied to safety.* MIT Press. https://direct.mit.edu/books/oa-monograph/2908/Engineering-a-Safer-WorldSystems-Thinking-Applied

Liang, R., Pan, W., Zuo, Q., Zhang, C., Chen, S., Chen, S., & Deng, L. (2025). Modeling visual fatigue in remote tower air traffic controllers: A multimodal physiological data-based approach. *Aerospace, 12*(6), 474. https://doi.org/10.3390/aerospace12060474

Manning, C. A., Mills, S. H., Fox, C., Pfleiderer, E. M., & Mogilka, H. J. (2002). *Using air traffic control taskload measures and communication events to predict subjective workload* (Report No. DOT/FAA/AM-02/4). Federal Aviation Administration, Office of Aerospace Medicine, Civil Aerospace Medical Institute.

Masi, G., Amprimo, G., Ferraris, C., & Priano, L. (2023). Stress and workload assessment in aviation—A narrative review. *Sensors, 23*(7), 3556. https://doi.org/10.3390/s23073556

Mitchell, L. (2023, October 3). *NASA studies human pilots to advance autonomous air taxis.* NASA Armstrong. https://www.nasa.gov/centers-and-facilities/armstrong/nasa-studies-human-pilots-to-advance-autonomous-air-taxis/

Monk, K. J., Fettrow, T. D., Galvan-Garza, R. C., & Kraft, A. E. (2024). *Physiological and subjective responses of pilots during AAM flight testing with automated systems* (NASA/TP-20240007669). NASA Ames Research Center. https://ntrs.nasa.gov/citations/20240007669

Muñoz-de-Escalona, E., Leva, M. C., Gianini, G., Lopez de Frutos, P., & Jadronova, M. (2025). Air traffic controllers communication analysis as a proxy of task demand and mental workload: Using voice recording markers for safety critical task. *Safety Science, 186*, 106928. https://doi.org/10.1016/j.ssci.2025.106928

National Aeronautics and Space Administration. (2022). *AOSP newsletter Oct–Dec 2022 Quarter 1.* https://www.nasa.gov/wp-content/uploads/2023/08/aosp-newsletter-oct-dec-2022-q1-v2.pdf

Neff, P. S. (2019). Crew resource management for large unmanned aircraft systems operations. *International Journal of Aviation, Aeronautics, and Aerospace, 6*(3). https://doi.org/10.58940/2374-6793.1303

Parasuraman, R., & Riley, V. (1997). Humans and automation: Use, misuse, disuse, abuse. *Human Factors, 39*(2), 230–253. https://doi.org/10.1518/001872097778543886

SESAR Joint Undertaking. (n.d.). *Initial U-space implementation in controlled and uncontrolled airspace* (Solution #0311 / Release 15).

SESAR Joint Undertaking. (2022). *Assessment of a performance-based surveillance function in the U-space* (Paper 110).

SESAR Joint Undertaking. (2023). U-Space concept of operations (4th ed.). https://www.sesarju.eu/sites/default/files/documents/reports/U-space%20CONOPS%204th%20edition.pdf

Shenzhen Government Online. (2018, November 21). *City pilots UAV traffic management program.*

Shenzhen Government Online. (2024). *Low-altitude economy (feature hub).* https://www.sz.gov.cn/en_szgov/news/infocus/LowAltitudeEconomy/LowAltitudeEconomyinFocus/

Tresslar, T. (2023, October 25). *AFWERX Prime, FAA agree to share flight-test information.* Air Force Research Laboratory Public Affairs. https://www.afrl.af.mil/News/Article-Display/Article/3561121/afwerx-prime-faa-agree-to-share-flight-test-information/

U.S. Department of Transportation, Office of Inspector General. (2021). *FAA has made progress in implementing ASIAS, but work remains to better measure program effectiveness* (Audit report).

U.S. DOT Volpe Center. (2024). *Up, Up, and Away: Innovations in Advanced Air Mobility (Thought Leadership Series).* https://rosap.ntl.bts.gov/view/dot/77200

Xu, H., Wang, L., Zou, J., Zhang, J., Li, R., Ma, X., Wang, Y., & Zhao, Y. (2025). Recognising and explaining mental workload using low-interference method by fusing speech, ECG and eye-tracking signals during simulated flight. *Ergonomics*, 68, 1–18. https://doi.org/10.1080/00140139.2025.2511877

Yang, J., Yang, H., Wu, Z., & Wu, X. (2023). Cognitive load assessment of air traffic controller based on SCNN-TransE network using speech data. *Aerospace*, *10*(7), 584. https://doi.org/10.3390/aerospace10070584

6 Respond
Effective Reaction

INTRODUCTION

On a cold January afternoon in 2009, an Airbus A320 lost both engines after striking a flock of geese and instantly became a powerless glider above New York City. With fewer than four minutes before impact, Captain Chesley "Sully" Sullenberger and First Officer Jeffrey Skiles drew on decades of practice, executed a ditching checklist most pilots never perform in real life, and set the crippled jet down on the icy Hudson River. All 155 passengers and crew survived (Fraher, 2011). What seemed like a miracle was, in fact, the manifestation of *operational resilience*: well-trained humans exploiting robust aircraft structures, intuitive procedures, and cooperative air traffic coordination to transform an unfolding disaster into a controlled outcome (Pariès, 2011).

A similar demonstration of resilience occurred in 2010 when Qantas Flight 32 suffered an uncontained engine failure that crippled 21 of 22 primary systems. Facing a "Christmas tree" of cascading warnings, the crew stabilized the damaged Airbus A380 and returned safely to Singapore, averting what investigators later called "a catastrophic loss of the aircraft" (Creedy, 2022, para. 5). These episodes bring to light a central lesson: while *anticipation* and *monitoring* buy time, it is the *respond* capability—the system's immediate, adaptive reaction—that often separates tragedy from triumph.

This lesson is especially pertinent to the emerging ecosystem of advanced air mobility (AAM). Fleets of automated eVTOL aircraft and drones promise quieter, cleaner, point-to-point travel, yet they will operate at unprecedented density and complexity. Analysts project that as the number of sorties increases rapidly, the absolute number of in-flight contingencies will rise accordingly (National Academies of Sciences, Engineering, and Medicine, 2020). Resilience in AAM, therefore, hinges on swift, layered response: onboard autonomy must detect anomalies within milliseconds; networked traffic-management services must reroute flows in seconds; and human supervisors—often overseeing multiple vehicles—must retain the authority and situational awareness to intervene when software reaches its boundary conditions. Early NASA simulations caution that these operators may possess less hands-on flight experience than today's aviators, yet they will be expected to manage novel failures under intense time pressure (Nemeth & Holbrook, 2021).

Embracing the *respond* mindset means accepting that not every disruption can be predicted, but every disruption *can* be met with a prepared, practiced, and graceful reply. As resilience engineers remind us, systems and crews must be "prepared, and prepared to be unprepared" (Hollnagel, Woods, & Leveson, 2006, p. 11). Aviation's culture of relentless learning converts every handled emergency into improved

DOI: 10.1201/9781003546306-8

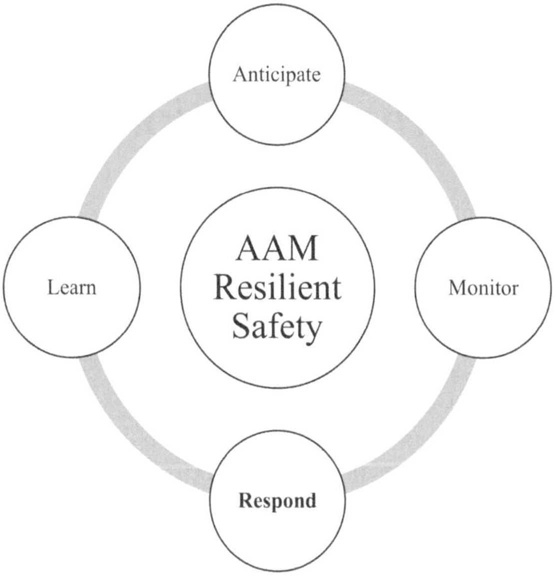

FIGURE 6.1 Respond function for resilient safety in AAM.

Note: Based on Hollnagel's (2009, 2015) four resilience abilities (anticipate, monitor, respond, learn).

training syllabi, updated algorithms, and smarter cockpit interfaces. By investing in these responsive layers—technological, organizational, and human—AAM stakeholders can ensure momentary anomalies do not cascade into catastrophes but are instead met with confidence, discipline, and creativity.

UNDERSTANDING THE RESPOND FUNCTION IN RESILIENT SAFETY

In resilience engineering theory, *respond* is one of four cornerstone capabilities that enable a system to be resilient, alongside the abilities to anticipate, monitor, and learn. *Responding* refers to an organization or system's capacity to react to what is happening (or has happened) in a timely and effective manner, thereby preventing or mitigating undesirable outcomes. In Erik Hollnagel's terms, responding is "knowing what to do," that is, being able to address regular and irregular variability, disturbances, or even opportunities by adjusting how things are done, activating prepared actions, or creating new solutions on the fly (Hollnagel, 2011; Hollnagel, 2015). In essence, a resilient AAM operation must *"know what to do, when and how to respond"* even under unanticipated conditions (Hollnagel, 2011, p. xx).

AAM presents a novel context for applying this *respond* capability. Traditional aviation systems (e.g., commercial airlines or rotorcraft operations) have well-established emergency procedures and trained crews to respond to in-flight issues. By contrast, AAM envisions highly automated vehicles (many of which are unmanned or minimally crewed), dense operations in urban airspace, and new stakeholders, such as fleet operators and automated air traffic services. These differences pose both challenges and opportunities for a resilient response. On one hand, AAM vehicles will rely heavily on automation and autonomy, potentially reacting faster than humans to certain hazards. On the other hand, limited human involvement (or novel roles like remote pilots) means the system must preserve "the potential for human successes" even as it minimizes opportunities for human error (Nemeth & Holbrook, 2021, p. 1). As Nemeth and Holbrook (2021) caution, simply removing humans may eliminate some errors, but it also risks losing the human adaptability and problem-solving that are vital for resilience. Therefore, designing a resilient response in AAM requires carefully balancing autonomous system responses with human performance capabilities.

The *respond* function in resilient AAM safety can be seen as both proactive and reactive. It is not only about reacting after a failure occurs, but also about being prepared to react swiftly and effectively *when* a disruption arises. Hollnagel emphasizes that no system can survive unless it can respond to what happens, which first requires detecting that something has occurred, recognizing the need for a response, and then knowing how to act (and when to start or stop the response) (Hollnagel, 2015). To achieve this, a system should ideally have pre-planned responses and resources at the ready, or be flexible enough to rapidly reconfigure itself to obtain the necessary resources. For AAM operations, this translates to measures such as established emergency procedures, automated fail-safe modes in vehicles, on-call human support, and communication protocols with external responders.

Another key aspect of responding is timeliness. Responses in AAM must occur quickly enough to be effective in a dynamic environment. For example, if an urban air taxi's propulsion fails, the time window to choose an emergency landing site is very small; the system (whether an onboard autonomy or a remote pilot) must react almost immediately to avoid a crash. In resilient performance terms, "responses must furthermore be both timely and effective so that they can bring about the desired outcome before it is too late" (Hollnagel, 2015, p. 6), emphasizing the need for continuous monitoring (to detect anomalies early) and decision support tools that expedite correct responses. A delayed or inappropriate response can negate any benefits of anticipation or monitoring. In practice, a timely response in AAM might mean automated hazard detection systems triggering maneuvers within milliseconds or well-drilled human operators initiating contingency procedures within seconds.

Finally, it is essential to note that *respond* in the resilient safety framework does not operate in isolation. It is closely tied to other resilience potentials. The ability to anticipate (foresee issues) and monitor (track system health and environment) provides the context that triggers a response. The ability to learn from past incidents improves future responses.

TECHNICAL RESPONSE MECHANISMS IN AAM SYSTEMS

A defining feature of AAM is the prominent role of technical systems—automation, autonomy, and AI—in managing flights. Many AAM concepts involve highly automated vehicles that may not carry a pilot on board (e.g., remotely piloted or fully autonomous air taxis and delivery drones). As such, the technical system must be capable of handling a wide range of contingencies in real-time while allowing automation and AI to enable responsive action through features such as autonomous contingency management, detect-and-avoid (DAA) systems, fail-safe designs, and intelligent decision support.

Automation for Autonomous Contingency Management

One of the most critical technical needs in AAM is autonomous contingency management—the ability of the vehicle and supporting infrastructure to detect anomalies or failures and take corrective action without needing immediate human intervention. A National Academies of Sciences, Engineering, and Medicine report (2020) found that due to the expected increase in traffic density and a likely decrease in the availability of highly trained human pilots, *"autonomy for contingency management will be an essential component of advanced aerial mobility"* (National Academies of Sciences, Engineering, and Medicine, 2020, p. 6). In other words, AAM vehicles should be equipped to handle emergencies quickly and restore a safe flight state, even in the absence of a highly qualified onboard flight crew (National Academies of Sciences, Engineering, and Medicine, 2020).

In traditional aircraft, a human pilot is the primary contingency manager (handling engine failures, bad-weather deviations, etc.). But in AAM, vehicles may be operated by remote pilots overseeing multiple aircraft or even be fully autonomous. Therefore, onboard systems are being designed with the authority to make firm decisions independently in emergency scenarios. For example, if an electric air taxi experiences a loss of communication link with its ground controller, it cannot wait for instructions—it must automatically execute a predefined safe procedure (such as hovering in place, ascending to a safe altitude, or returning to a designated landing site). Indeed, lost-link scenarios are a classic case where the vehicle's automation must respond immediately. As one NASA study notes, "*Lost communications must be dealt with directly in certification to ensure the onboard automation can safely manage the vehicle's mission and flight trajectory without any human assistance during the outage*" (Wing et al., 2020, p. 2). In practice, a responsive AAM autopilot would detect the communication loss (*the trigger*), consult its programmed contingency plan (the decision logic), and then autonomously navigate to the nearest safe landing site or loiter at a safe altitude and area until the link is restored (*the action*).

Autonomous response capabilities also include DAA systems to prevent mid-air collisions. DAA technology utilizes sensors (such as radar, LiDAR, and cameras) and algorithms to detect other aircraft or obstacles and maneuver to avoid conflict. This is essentially a real-time response mechanism to the hazard of traffic conflicts. Given that AAM operations in cities will likely encounter numerous potential conflicts (with other drones, helicopters, birds, and buildings), robust DAA automation is necessary. The system must continuously monitor the environment and, if another object is on a collision course, immediately alter the flight path without waiting for a human command. Many AAM prototypes incorporate *sense and avoid* or collision-avoidance systems, often drawing on research from unmanned aircraft. These automated maneuvers are examples of protective responses that happen faster than a human could react, thereby enhancing safety.

However, designing autonomous responses is a challenging task. The National Academies of Sciences, Engineering, and Medicine (2020) cautions that contingency automation must be highly reliable and well-structured—it should "*select and execute mitigation actions accurately and without loss of control or unnecessary disruption*" (p. 7). For instance, if an AAM drone detects a critical battery failure, an autonomous response might be to perform an emergency landing in a pre-approved area. This action must be done *gracefully*: the aircraft should avoid crashing into obstacles, notify the broader network of its intentions (so other traffic and the management system are aware), and minimize risk to people on the ground. Achieving this requires extensive testing, sensor fusion, and perhaps redundancy in decision logic to avoid false triggers. It also requires that the criteria for triggering such autonomy are well-defined—e.g., the system should neither miss a real emergency (false negative) nor trigger unnecessarily for a benign anomaly (false positive).

Early AAM deployments will likely start with simpler contingency automations based on encoding well-understood procedures. One finding of the National Academies study was that "*encoding well-established contingency management procedures into autonomy will provide a rich baseline capability for automated contingency management in the near term*" (National Academies of Sciences, Engineering, and Medicine, 2020, p. 6). In practice, this means taking the checklists and emergency actions that human pilots use (for engine failure, loss of GPS, etc.) and programming the autonomous system to follow those steps automatically. Over time, with the advancement of machine learning and the accumulation of more operational data, autonomy may evolve into more sophisticated responses. But initially, this approach ensures the automated behavior is aligned with proven safety practices (which can then be certified using existing methods). For example, many current drones have a return to home feature that automatically activates if communication or GPS signals are lost, or when reaching the limits of a geofenced area—this is a straightforward translation of a contingency procedure into code.

System Redundancy and Fail-Safe Design for Resilient Response

Beyond software intelligence, physical design choices in AAM vehicles contribute enormously to their ability to respond safely to failures. The principle of redundancy—having backup systems to take over if one fails—is a cornerstone of aerospace safety, and it is being applied aggressively in eVTOL and drone designs. Many AAM aircraft feature multiple independent rotors, motors, or power sources so that a single failure does not lead to a catastrophic outcome. For example, the Volocopter air taxi uses 18 rotors for lift; it can tolerate one or even several rotors malfunctioning and still remain airborne long enough to land safely (Ackerman, 2021). According to the manufacturer, "*Volocopter ensures safety by redundancy. The most visible of that is our rotors*," with all critical systems duplicated or backed up (Volocopter, n.d.). This multi-redundant architecture enables the vehicle to respond to a component failure by redistributing thrust to the remaining propellers—a response inherent in the design. Indeed, in test flights, the Volocopter demonstrated its ability to land safely even if some of its rotor drives failed (Cohen, 2013). Other eVTOL companies adopt similar approaches: distributed electric propulsion inherently provides redundancy (multiple small motors vs. one big engine) and is often combined with backup flight computers and multiple batteries. The goal is that the failure of any single component will not critically impair the vehicle's ability to respond and maintain control.

Another fail-safe measure in some AAM designs is the inclusion of ballistic parachutes or deployable airbags. In Dubai's autonomous air taxi trials, for instance, the Volocopter prototype was fitted with an optional whole-aircraft parachute as well as nine independent battery systems (Ong, 2017). The parachute could deploy to slow the vehicle's descent and prevent a free-fall crash in the event of a severe emergency. While a parachute is a crude response (essentially yielding fine control but minimizing impact energy), it exemplifies how engineers are building in layered responses—first relying on automation to handle failures, but if all else fails, having a passive safety device to protect occupants and bystanders. Similarly, some heavy-lift delivery drones are equipped with parachutes; the U.S. FAA has issued waivers for beyond visual line of sight (BVLOS) drone operations, provided that an approved parachute recovery system is in use (Federal Aviation Administration, 2022). In practice, Zipline's fixed-wing medical delivery drones automatically deploy a parachute if they lose control or encounter a serious malfunction, thus responding to any anomaly by safely descending into a predefined drop zone. This strategy has been effective in thousands of flights—Zipline estimates that the emergency parachute deploys in roughly one out of every thousand flights as a last-resort safety response (Ackerman & Koziol, 2019). Those rare deployments have significantly reduced the risk to people on the ground, enabling regulators to permit the rapid expansion of drone delivery services while maintaining safety.

Energy and propulsion systems in AAM vehicles are also being designed with responsiveness in mind. For electrically propelled aircraft, battery management systems can isolate failing battery cells to prevent thermal runaway from spreading, effectively responding to an internal fire risk by cutting off or rerouting power around the affected module. Some conceptual designs even consider jettisoning an overheated battery module in mid-air if it is at risk of fire (a drastic measure, likely only feasible over uninhabited areas). Hybrid eVTOLs (if used) may have multiple engines or an engine-plus-battery combination, allowing one to power the vehicle if the other fails. The European Union Aviation Safety Agency (EASA) has codified these expectations in its Special Condition VTOL standards: in the *Enhanced* category of eVTOL (intended for passenger transport over urban areas), *"it should always be possible to continue safe flight or land after any single failure, without excessive workload"* on the crew (EASA, 2019, p. 5). In essence, the aircraft must be tolerant to at least one failure, which typically means redundancy or alternate modes, and the response to that failure (e.g., transitioning to a backup system or flight mode) must not overwhelm the pilot. This is a regulatory embodiment of a resilient response: the vehicle itself should handle failures gracefully, so that a single failure does not lead to the loss of the aircraft.

Flight control systems are another focus. Most eVTOLs use digital fly-by-wire control, which offers opportunities for built-in envelope protection and stability augmentation. Fly-by-wire can automatically compensate for disturbances (like wind gusts or even a rotor failure) faster than a human could react. EASA's approach requires that eVTOL flight control systems include flight mode annunciations and envelope protections so that the aircraft will autonomously prevent the pilot (or autopilot) from commanding manoeuvres that exceed structural or aerodynamic limits (Airbus, 2023; EASA, 2021) In practice, this means the control software will respond to dangerous control inputs by modifying or vetoing them—for example, if a pilot or autonomous flight computer tries to climb too aggressively and risk an aerodynamic stall, the system might automatically limit the climb rate. This type of protective response ensures that even if a pilot is inexperienced or an automation glitch occurs, the vehicle remains in safe flight conditions. Envelope protection can be seen as a form of *pre-emptive response*: it does not wait for an accident to happen; it continuously responds to keep the flight within safe bounds.

HUMAN PERFORMANCE AND RESPONSE IN AAM OPERATIONS

While advanced automation is a hallmark of AAM, human performance and decision-making remain critical to resilient operations. Humans—whether they are onboard pilots, remote operators, maintenance crews, or air traffic managers—bring adaptability, creativity, emotional intelligence, and ethical judgment that purely technical systems lack. However, AAM introduces new human factors challenges: operators may be overseeing multiple vehicles, pilots might have reduced training (due to highly automated "simplified" controls), and traditional roles (like air traffic control) will evolve with new procedures. Therefore, we need to examine the human element of the *respond* function in the following aspects:

The Role of Pilots and Operators in Emergency Response

In early AAM implementations, human pilots will likely still be involved in some capacity for passenger-carrying flights. NASA's roadmap anticipates a gradual transition: initially, certified pilots will be on board eVTOL air taxis; over time, operations may shift to remote supervision and eventually full autonomy (Wing et al., 2020). During this evolution, the nature of the pilot's role changes, but their responsibility to respond to anomalies remains vital. With an onboard pilot, responding in AAM is not fundamentally different from conventional aviation—the pilot monitors the aircraft and environment. If a malfunction or hazard arises, they execute appropriate procedures (with help from automation as needed). However, AAM aircraft may simplify some tasks via automation (for example, auto-stabilization or automated landing sequences), potentially reducing pilot workload during routine operations. The paradox is that, in emergencies, the pilot may need to step in *precisely when automation reaches its limits*—yet the pilot may be relatively *out of the loop* due to the high level of normal automation. This is often referred to as the *out of the loop* problem in automation: humans can lose situational awareness when monitoring a highly autonomous system, resulting in slower or less informed responses when they suddenly need to take over. Thus, training and cockpit design must ensure the pilot can re-engage quickly and effectively when an alert or unusual situation occurs.

In remote operations, a human operator may oversee several vehicles simultaneously from a control center. The concept of a remote supervisor operator (RSO)—sometimes described as an "*m:n*" supervision model, where one operator manages *m* vehicles—is being considered for AAM to improve efficiency (Wing et al., 2020). Here, the human's role is more of a manager or supervisor: they intervene only when automation requests help or when non-routine scenarios occur. The challenge is that one person could be responsible for multiple aircraft, so their ability to respond to an emergency in any single vehicle could be delayed or limited by concurrent demands. If two drones experience issues simultaneously, the operator might face a task saturation problem. Therefore, AAM operational concepts usually limit the ratio of vehicles to operators under normal conditions and require automated fail-safes to handle incidents when the operator is overloaded or unavailable. For example, an RSO might normally handle up to 10 drones. However, if an emergency arises with one, the system might automatically reduce the workload (perhaps by pausing non-urgent tasks or transferring other vehicles to another supervisor) so the operator can focus on the incident.

It is widely recognized that workload and situational awareness are key determinants of human response effectiveness. Studies have shown that increasing automation and traffic density can lead to an unmanageable workload for human controllers if not mitigated. For example, Edwards et al. (2019) conducted human-in-the-loop simulations for urban air mobility air traffic control. They found that at medium and high densities of UAM traffic, using traditional communication and control methods led to excessive controller workload, causing them to refuse or delay additional traffic. However, when specific adjustments were introduced, such as implementing structured routes and reducing verbal communications via automated data exchanges, workload was significantly reduced and performance improved (Edwards et al., 2019). This indicates that to maintain human responsiveness, the operational environment must be *designed for human manageability*. High workloads or complexity can severely impair a human operator's ability to detect problems and respond correctly. In AAM operations, this may involve providing decision support tools to the operator, limiting the number of vehicles per operator until AI can reliably take over more tasks, and simplifying user interfaces to highlight the most critical information.

Human response also relies on training and experience. Because AAM is a developing concept, there is currently a limited pool of experienced AAM pilots and operators. In fact, one driver for autonomy in AAM is the scarcity of pilots and the high cost of training them. Concepts like simplified vehicle operations aim to reduce the skill required by automating much of the flight so that pilots with less experience (or new types of operators) can manage the vehicle (Wing et al., 2020). This raises an important consideration: a resilient response might paradoxically become *harder* if pilots are less trained or experienced in handling emergencies. The National Academies of Sciences, Engineering, and Medicine (2020) report explicitly warns that even well-trained traditional pilots struggle with the high workloads typical in emergencies requiring contingency response, so pilots with substantially less training and experience will likely be even less prepared for such situations. To counter this, the system's design and training programs must be adjusted accordingly. For instance, more simulator-based emergency training could be required for AAM remote operators, even if day-to-day flying is largely automated. Additionally, interfaces can be designed to guide less experienced operators through emergency steps (e.g., a checklist that appears during a malfunction and highlights recommended actions on the ground station console).

Despite high automation, humans are expected to remain the ultimate decision-makers for certain critical responses, especially those involving ethical judgments or complex trade-offs. For example, if an air taxi's propulsion fails over a populated area, an autonomous system might have options: attempt to glide toward a crowded park or ditch in a river. A human might be better suited to assess the least-worst consequences in that scenario, weighing passenger safety against bystander risk. Ensuring a human can intervene or at least oversee such decisions is often desirable for accountability and public acceptance. Therefore, AAM operations centers will likely have supervisors who can take manual control or send override commands in dire situations. In practice, a remote pilot might have a manual joystick or command terminal ready, even if 99% of flights run autonomously, so that in the 1% of cases where human judgment is needed, they can instantly step in.

It is also worth noting that the concept of crew resource management (CRM) extends to AAM but in new forms. In a traditional aircraft cockpit, CRM principles enhance the coordination between a pilot and a co-pilot in responding to emergencies. In AAM, the "crew" might consist of a single remote operator and an AI assistant, or a network of support personnel (e.g., safety observers, maintenance engineers providing advice during a technical issue, or local emergency services on call). Developing effective teamwork protocols between all humans involved is crucial. If, for example, a drone is forced to land in an unplanned location, the operations center should swiftly coordinate with field personnel to recover it and, if necessary, with local authorities. These coordinated response actions require clear procedures and communication channels established in advance. Essentially, everyone—the pilot/operator, the supporting team, and external partners—should be aware of their role in the event of an emergency to prevent confusion and delays.

Human–Technology Interaction Under Stress

The interface between humans and the automated AAM system is a crucial factor in determining the success of responses when pressure is high. A poorly designed human-computer interface can mislead or overload an operator. In contrast, a well-designed one can significantly enhance human response by providing the correct information at the right time. Under emergency conditions (such as stress, time pressure, or high stakes), the cognitive load on the human operator increases significantly, and the likelihood of error also rises. Thus, human factors engineering for AAM focuses on creating interfaces and support tools that reduce cognitive load, prevent confusion, and guide effective action during contingencies.

One key principle is *salient alerting*, which allows the system to effectively gain the operator's attention for critical events without being too frequent. This means tuning the alert systems (visual warnings, audible alarms, and perhaps haptic feedback) to differentiate levels of urgency and to provide as much information as possible. For example, a mild engine performance deviation might prompt a caution message that the operator can acknowledge at their convenience. In contrast, a complete propulsion failure should trigger a loud, distinctive alarm and a clear on-screen directive that immediate action is needed. However, designers must be careful to avoid *excessive* alarms—if an AAM cockpit or control station bombards the operator with frequent warnings (especially if many are false or nuisance alerts), the operator may become desensitized or confused about priorities. Striking the right balance in alerting—filtering sensor data so that only actionable issues produce alerts, and designing those alerts to convey the nature of the problem and the suggested response—is critical. This balance directly affects how quickly and accurately a human operator will react. Studies of conventional flight decks show that an excess of auditory warnings can lead pilots to overlook the single, truly critical alarm (Dehais et al., 2014; Edworthy et al., 1995; Sarter & Woods, 1995). AAM developers should therefore adopt *adaptive alerting* strategies that escalate warnings appropriately and suppress lower-priority notifications when a high-priority emergency is in progress, as recommended in contemporary flight-crew alerting guidance (Federal Aviation Administration, 2011).

Another important concept is mode awareness. Modern aircraft automation often features multiple modes (e.g., different autopilot modes or flight control laws), which can cause confusion among pilots regarding the system's current mode and its intended actions—a factor that has contributed to past accidents. For AAM, where the automation authority will be even higher, ensuring the human knows the current mode/state of the system is vital. EASA's guidance for eVTOL (Enhanced category) explicitly states "the crew should be aware of the flight mode," especially when automation degrades or switches modes (EASA, 2021). If, for instance, the flight control system switches to a backup control law after a failure, the pilot/operator interface should clearly indicate this (through mode annunciation and possibly an advisory on limitations), so the human operator understands any new constraints or required actions. Loss of mode awareness can lead to erroneous responses (for instance, if an operator thinks the vehicle is holding position when it is actually in a return-to-base mode, they might give incorrect commands). Therefore, interfaces must display the active mode of automation and any changes to it, especially during abnormal events.

Furthermore, *decision support displays* can significantly aid human responders. For instance, if an emergency landing is needed, a smart display could instantly show the operator the nearest potential landing sites (vertiports or safe open areas), taking into account the current altitude, vehicle glide capability, wind, and population density below. Instead of the pilot mentally figuring out where to land, the system presents a short list of viable options, each with an indication of estimated safety or distance. This type of tool simplifies a complex decision, making it a more straightforward choice, and thereby accelerates *the response*. Research prototypes exist for such displays, often integrating map data, vehicle performance models, and even crowd density maps to suggest a landing spot that minimizes risk to people on the ground. Similarly, for a remote supervisor managing multiple drones, if one goes into a contingency mode, the interface might automatically bring that drone's status to the forefront and provide a *snapshot* of key info (e.g., *"Drone #5—lost GPS signal, now navigating via backup INS, returning to base"*). This frees the human from having to dig through data or manually take control immediately, allowing them to monitor whether the automated response is working and only intervene if necessary. Companies like Wing have implemented features that allow an operator to, with one command, instruct all drones in an area to pause takeoffs, proceed with contingency landing or pause deliveries in a service area, if a sudden hazard is reported—an interface design that enables a *collective response* with minimal human effort (Wing Team, 2024).

A design approach that treats humans and automation as a collaborative team is known as human-autonomy teaming (HAT). Under high-stress conditions, a well-orchestrated HAT means the automation handles what it is best at (fast, low-level control actions and continuous monitoring of dozens of parameters). At the same time, human does what they are best at (contextual judgment and creative problem-solving for novel scenarios). For example, during a severe weather encounter, the automation might keep the vehicle stable and immediately adjust the speed/altitude to counteract turbulence. At the same time, the human pilot considers *higher-level decisions*, such as whether to divert the flight to a different destination. A hallmark of good HAT is transparency—the automation should make its intentions and reasoning understandable to humans. In practical terms, this could mean an AI copilot that provides verbal or textual explanations, such as "*Wind gust detected, initiating heading change to maintain stability*," so the pilot is not caught off guard by the vehicle suddenly turning, and also understands the rationale behind it. *Transparency* builds *trust*, which is critical; if operators trust the automation, they are more likely to let it handle things and only step in when truly necessary (preventing the knee-jerk overrides that sometimes cause incidents). Conversely, if the automation acts opaquely, the human might either intervene too often (disrupting the response) or not intervene when they should (because they assume the automation has it under control).

Ultimately, under stress, psychological and ergonomic factors become significant. Stress can cause tunnel vision, auditory exclusion, and impaired decision-making. Control centers for AAM should be *ergonomically arranged* to minimize fatigue and stress—e.g., comfortable seating, uncluttered displays, and minimal unnecessary distractions. In cockpit design for eVTOLs, manufacturers emphasize simplicity and intuitiveness to accommodate pilots who may not have airline-level training. For example, some eVTOLs use simplified controls (like a single joystick and a tablet-like display) instead of the myriad dials and switches in a commercial airliner. The intent is that, in an emergency, the pilot is not overwhelmed by a dizzying array of instruments or a complex series of steps; instead, critical information is prioritized on a central display, and controls respond in a straightforward manner. One could imagine an eVTOL emergency checklist being partially automated—when a failure is detected, the system might automatically perform initial steps (e.g., isolating an overheating battery) and present the pilot with a clear prompt for the next step, e.g., "*Land at nearest site within 2 minutes: Yes/No?*"

MODELING AND EVALUATING RESPONSE CAPABILITIES

To ensure that both technical systems and humans in AAM will respond effectively to various situations, it is valuable to model and analyze the *respond* mechanisms. One approach from resilience engineering is the functional resonance analysis method (FRAM), which models complex socio-technical systems in terms of functions and their interactions (Hollnagel, 2012). In this section, we discuss how FRAM can be applied to AAM operations to understand responsiveness, as well as other frameworks like Hollnagel's resilience analysis grid (RAG) that help evaluate an organization's *respond* potential.

Applying FRAM to Advanced Air Mobility

Erik Hollnagel developed FRAM as a method to analyze how work is performed in complex systems, recognizing that outcomes emerge from non-linear interactions (or "resonance") of functions, especially when each function can vary in performance (Hollnagel, 2012). Unlike linear cause-and-effect models, FRAM does not separate failures from normal performance; instead, it examines how everyday performance variability can combine in unexpected ways to produce both successful and unsuccessful outcomes. This perspective is particularly apt for AAM, where numerous components (software, hardware, humans, and environment) interact in real-time, and unexpected couplings can occur.

In FRAM, the system is described in terms of functions (which can be technical functions, human tasks, or organizational activities). Each function is characterized by six aspects: *input*, *output*, *precondition*, *resource*, *control*, and *time*. These aspects define what each function requires and produces, as well as how it interacts with others. An output of one function can serve as an input or a control for another, thus coupling them. Variability in one function's performance can spread to others through these couplings—this is what Hollnagel calls *functional resonance*.

To illustrate, in Figure 6.2, a simplified work-as-imagined (WAI) FRAM model of an AAM operation during a contingency can be considered for, say, an unexpected weather hazard encountered during flight:

Function 1: Set Baseline & Ready the Operation—publish the basic rules, tools, and timing before flight.
Output: Operational policies & thresholds Information & tools ready; Time windows & update intervals.
Input: Post-event update (from Function 9).
Resource: Updated policies & thresholds (from Function 9).
Function 2: Detect Weather Hazard—onboard/external data indicate a windshear or storm cell.
Output: Hazard alert Current weather picture.
Input: Information & tools ready (from Function 1).
Resource: Information & tools ready.
Control: Operational policies & thresholds.
Time: Time windows & update intervals.
Function 3: Decide on Contingency—choose to continue, divert, or land.
Output: Contingency decision.
Input: Hazard alert (from Function 2).
Precondition: Current weather picture (from Function 2).
Control: Operational policies & thresholds.
Time: Time windows & update intervals.

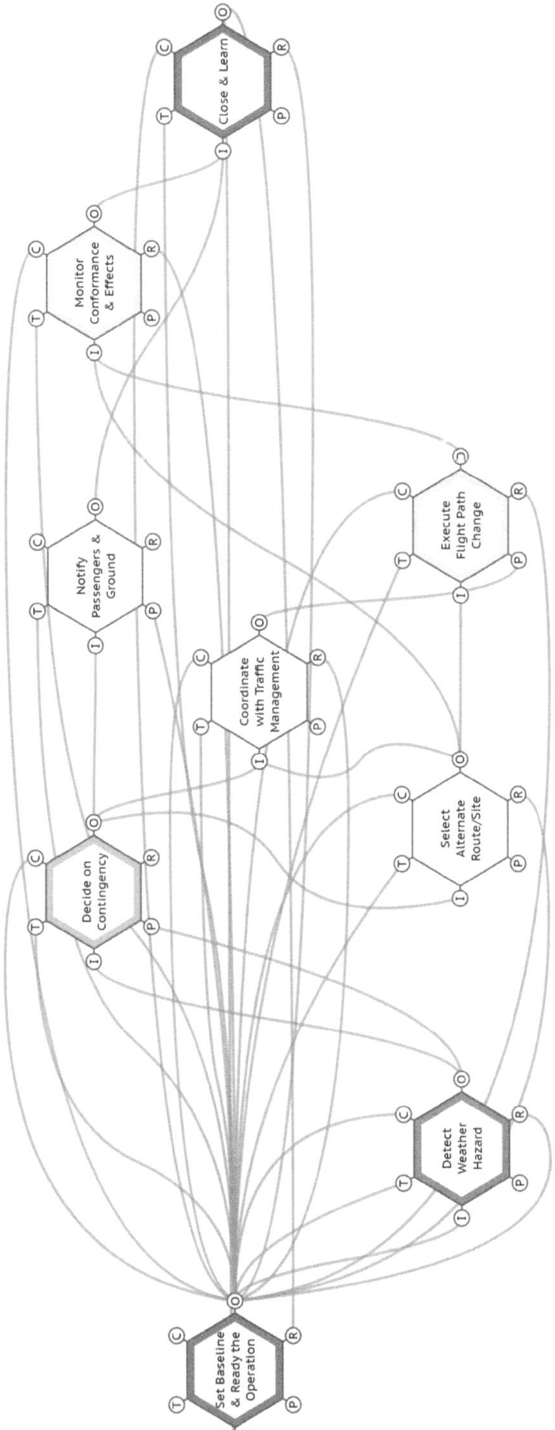

FIGURE 6.2 FRAM model for weather hazard, work as imagined (WAI). (Credit: Author; created with FRAM Model Visualizer.)

Function 4: Select Alternate Route/Site—pick a safer route or landing site.
Output: New flight plan.
Input: Contingency decision (from Function 3).
Resource: Information & tools ready.
Control: Operational policies & thresholds.
Time: Time windows & update intervals.

Function 5: Coordinate with Traffic Management—notify UTM/ATC and get a response.
Output: Network acknowledgement & constraints.
Input: Contingency decision or New flight plan (from Functions 3/4).
Resource: Information & tools ready.
Control: Operational policies & thresholds.
Time: Time windows & update intervals.

Function 6: Execute Flight Path Change—fly the new trajectory.
Output: Aircraft status update.
Input: New flight plan (from Function 4).
Precondition: Network acknowledgement & constraints (from Function 5).
Resource: Information & tools ready.
Control: Operational policies & thresholds.
Time: Time windows & update intervals.

Function 7: Monitor Conformance & Effects—watch adherence and any impacts.
Output: Conformance status & advisories.
Input: Aircraft status update (from Function 6) and New flight plan (from Function 4).
Resource: Information & tools ready.
Control: Operational policies & thresholds.
Time: Time windows & update intervals.

Function 8: Notify Passengers & Ground—inform occupants and ground teams.
Output: Stakeholders notified.
Input: Contingency decision (from Function 3).
Precondition: Information & tools ready.
Time: Time windows & update intervals.

Function 9: Close & Learn—close the event and update the baseline.
Output: Post-event update Updated policies & thresholds.
Input: Conformance status & advisories (from Function 7) and Stakeholders notified (from Function 8).
Resource: Information & tools ready.
Control: Operational policies & thresholds.
Time: Time windows & update intervals.

By mapping this out, FRAM enables us to visualize how a response to the weather hazard emerges from the interaction of its constituent functions. We can then ask: what happens if one function varies from its nominal performance? For example, what if the *Detect Weather Hazard* function is delayed or produces a *false negative* (perhaps the sensor failed to detect the windshear in time)? The model would suggest that the *Decide on Contingency* function might then proceed without crucial input, potentially leading the system to fail to respond to the hazard or respond too late. Or, if *Coordinate with Traffic Management* fails (maybe the datalink was down, so traffic managers do not know the drone has diverted), there could be downstream effects like a potential conflict with another aircraft because traffic management did not clear the new route.

The FRAM approach encourages consideration of performance variability, as humans may take longer to decide under stress, sensors may exhibit noise or lag, and algorithms may have conservative or aggressive tuning. Instead of treating each potential failure in isolation, FRAM examines how these variations in each function can interact with one another. While it could be used prospectively to identify hazards, examine feasibility of mitigation strategies, or simply help us comprehend the activity itself (Functional Resonance Analysis Method, n.d.). It is not strictly a risk assessment tool with probabilities; instead, it creates a realistic model of work that can underpin various qualitative and semi-quantitative analyses.

Researchers have begun applying FRAM to AAM and drone operations. For example, Fowler et al. (2024) conducted a FRAM analysis of an eVTOL operational concept in urban centers, identifying challenges and risks in how different functions (such as air traffic services, vehicle control, and ground support) might interact. Their preliminary findings suggest that interdependencies between functions, such as those between a remote pilot's actions and the vehicle's automated systems, can create complex outcomes that are not evident from a linear analysis. Another study applying FRAM from a joint cognitive systems perspective argues the method is well-suited to represent distributed human–technology work and, by extension, future human–AI teaming in AAM (Adriaensen et al., 2019). A broad review also shows aviation remains a dominant FRAM-application domain and that prospective FRAM use has grown markedly, supporting its relevance for emerging aviation concepts (Patriarca et al., 2020).

Conducting a FRAM analysis for an AAM operation typically involves: (1) identifying the key functions in the scenario, (2) characterizing each function with as many of the six aspects as relevant, (3) checking the model's completeness and consistency, (4) assessing potential variability in each function's performance (e.g., what if this function is delayed? what if its output is low quality or intermittent?), and (5) looking at how those variabilities could spread or amplify through couplings—the functional resonance. Step (5) essentially asks: *Could multiple small variabilities align to produce a large, unsafe outcome?* If so, those are points where either the design should add "dampers" (to reduce variability) or responders should get additional support or training.

By applying FRAM to, for example, an autonomous air taxi emergency landing process, safety engineers might discover that two functions—perhaps "*detect obstacle at landing site*" and "*communicate decision to control center*"—are both dependent on a common resource (the vehicle's primary computer). This coupling means that if that computer is overwhelmed or fails, both functions could underperform simultaneously, possibly leading to an unsafe landing without clearance or obstacle avoidance. Recognizing that, engineers could add a backup processor for obstacle detection or simplify the communication function during emergencies to reduce the load. In this way, FRAM helps design robustness into the *respond* capability by revealing *non-obvious connections* and *failure combinations*.

Evaluating The Respond Function's Potential: From FRAM to Resilience Metrics

While FRAM provides a qualitative model of how responses unfold, other frameworks aim to assess an organization's resilience capacities in a more structured way. One such tool is the RAG, introduced by Hollnagel, which evaluates the four resilience potentials (anticipating, monitoring, responding, and learning) via guided questions or checkpoints (Hollnagel, 2015). The RAG breaks down each potential (such as *respond*) into key areas and prompts an organization to consider how well it meets each one. By answering these, one can identify strengths and weaknesses in the current capability to *respond*.

For example, Table 6.1 below (adapted from Hollnagel, 2015) shows some key issues and example questions relating to the ability to *respond*:

Considering these factors in an AAM context, an operator might list events such as *GPS signal loss*, *vertiport blockage*, and *mid-air vehicle failure*, and ensure contingency plans exist for each (*Event list*). The *threshold* question is crucial to avoid hesitation or premature action—for instance, at what point does an AAM operations center declare an emergency and activate its incident response team? Is it when a vehicle has been unresponsive for 30 seconds, or only after it actually crashes? Clear

TABLE 6.1
Key Considerations for Response Capability

Issue	Key Questions (Examples)
Event list	What are the events/scenarios for which the system has a prepared response? How were these events selected (experience, regulations, risk assessment)? Who maintains this list?
Threshold	When is a response *activated*? What is the triggering criterion or threshold for initiating a response? Is it absolute or context-dependent?
Response options	What predefined response actions exist for each scenario? How was it determined that these responses are adequate (analysis, simulation, past data)?
Speed & duration	How quickly can a full response be mobilized? How long can the response be sustained? What is the minimum effective response level, and can it be maintained long enough?
Resources & capability	What resources (people, equipment, backup systems) are allocated for response readiness? Are some reserved exclusively for emergencies? Who ensures they remain ready?
Stop rules	What criteria determine when to terminate the *response* and return to normal operation? (For example, how to decide an emergency is over or contained.)
Verification	How is the readiness to *respond* verified and maintained (e.g., drills, equipment checks, training refreshers)?

Source: Adapted from Hollnagel's RAG approach.

triggers (possibly automated) make responses more timely. For *response options*, an AAM operator would catalogue actions such as: if propulsion is lost, deploy the auto-parachute and broadcast distress; if high winds are encountered at the vertiport, divert to the backup site; if communications are lost, enter a predefined holding pattern, etc. They must regularly evaluate if these options remain appropriate as technology and contexts evolve.

The *speed and duration* aspects are tied to resource planning. If we expect that a downed air taxi in a city will have first responders at the site within 5 minutes, do we have the people and assets positioned to achieve that? How long can those responders sustain their operations if the situation is prolonged (e.g., a drone on a power line needing hours to safely remove)? For AAM, especially in urban settings, the initial response may be swift (city fire departments can respond in minutes); however, in remote or high-density scenarios, backup plans are necessary if the immediate response stalls. The *resources & capability* issue asks if the organization has set aside the necessary tools: e.g., do they have spare communication links, a standby pilot who can take over, a recovery team for downed vehicles, etc.

Using a RAG or similar assessment, organizations can *rate themselves* or identify action items. For example, a gap might be found in *verification*: perhaps a company realizes it has procedures on paper, but has never conducted a full-scale drill of a drone crash scenario. That would prompt them to organize an exercise (for instance, simulating a drone emergency landing in a public area and testing how quickly the team responds and communicates with authorities).

Beyond FRAM and RAG, other safety-modeling techniques can also contribute to evaluating responsiveness. Techniques such as STAMP/STPA (systems-theoretic accident model and processes/system-theoretic process analysis) analyze control structures and can identify unsafe control actions that frequently stem from faulty or delayed responses (Leveson, 2012; Thomas & Leveson, 2013). For instance, an STPA study might flag that *"if the remote pilot does not issue a take-over command when the automation fails to avoid a collision, a loss can occur,"* thereby highlighting both the necessary response action and the operational context in which it could be missed (Thomas & Leveson, 2013). Furthermore, complementary simulations and stress tests—ranging from Monte Carlo scenario runs to human-in-the-loop exercises—provide empirical evidence of responsiveness by measuring how often a system successfully handles a given emergency, the time required, and any emerging bottlenecks (Bagheri et al., 2021; National Aeronautics and Space Administration, Ames Research Center, 2020).

One promising area is the development of *resilience metrics*. While safety has long used metrics like failure rates or incident counts, resilience metrics are more about capacities and performance under stress—e.g., *response time distribution* (how quickly the system can respond to significant anomalies), *recovery ability* (how well the system restores normal performance after a disturbance), and *coverage* (the variety of scenarios for which there is a response plan or capability). Researchers have used network analysis or multi-criteria methods to propose indices for resilient engineering factors (Azadeh et al., 2014, 2017). For an AAM operator, a simple metric might be the percentage of flights that experienced a contingency where the outcome was still safe. If 100 flights had some off-nominal event and 99 landed safely, that is a measure of effective response (perhaps to be improved until it is 100 out of 100). Other measures could include the *maximal response time* observed in drills versus what is required, or the *number of unplanned landings recovered without injury* as operations scale up.

In practical terms, evaluating responsiveness in AAM may involve conducting scenario-based analyses—using tools such as FRAM (Hollnagel, 2012) or STPA (Thomas & Leveson, 2013)—to verify that all identified hazards have viable responses. Operators can then conduct tabletop or live drills of emergency scenarios, recording completion times and lessons learned, as recommended in the aviation *Safety Management Manual* (International Civil Aviation Organization [ICAO], 2018). Continuous monitoring of real-world operations further "closes the loop" by capturing how incidents are actually managed and feeding that data back into training and procedures (National Academies of Sciences, Engineering, and Medicine, 2020). Finally, organizations should benchmark their performance against regulatory requirements and best practices—such as EASA's stipulation that pilot workload must remain acceptable even after system failures—to ensure that resilience is demonstrably adequate (European Union Aviation Safety Agency, 2019). Scenario-based simulator sessions that vary failure modes and track workload provide one practical way to meet this benchmark and to quantify system responsiveness.

Table 6.2 provides examples of contingency procedures, while Table 6.3 summarizes emergency procedures that may be encountered during a UAS flight.

TABLE 6.2
Examples of Response during UAS Contingency Procedures

#	Contingency Condition	Possible Failure	Crew	Item	#	Action
1	Accelerometer Failure	Unstable flight behavior	Pilot	Stability Sensors	A	Switch to manual control.
					B	Safely land the drone.
2	Appearance of Manned Aircraft	Airspace Conflict	GSO / Ground Staff		A	Call Out: "*Unknown Aircraft!*"
			Airspace Safety		B	Initiate landing
			Emergency Reporting		C	Report per ERP template
3	Appearance of Uninvolved UAS	Airspace Breach	GSO / Ground Staff		A	Call Out: "*Unknown UAS!*"
			Airspace Safety		B	Initiate landing
			Resume Operations		C	Ensure no simultaneous UAS operation before resuming.

TABLE 6.2 (*Continued*)
Examples of Response during UAS Contingency Procedures

#	Contingency Condition	Possible Failure	Crew	Item	#	Action
4	Bad Calibration	Drone circles or shakes	Pilot	Sensor Calibration	A	Use manual override.
					B	Safely land the drone.
5	Contingency Maneuver Lateral	Lateral Movement	GSO		A	Takeover manual control
			Lateral Path Correction		B	Stop lateral movement and return to flight geography.
			Persistent Lateral Deviation		C	Initiate termination (FTS)
6	Contingency Maneuver Vertical	Vertical Movement	GSO		A	Takeover manual control
			Vertical Path Correction		B	Stop vertical movement and return to flight geography.
			Persistent Vertical Deviation		C	Initiate termination (FTS)
7	Critical Battery	Low battery level	GSO	Battery Status	A	Monitor battery status.
					B	Initiate aircraft auto return if battery is critical.
8	Data Link Loss	Wi-Fi connection lost	GSO	Communication Link	A	Allow UAS to continue with preloaded flight paths.
9	Generator or Power Failure	Power outage in GCS or equipment	GSO	Power Supply	A	Ensure UPS and battery backups are active.
					B	No action needed if system unaffected.
10	Loss of Ground Control Computer	Ground Control Computer Failure	GSO		A	Call Out: "*Ground Control Computer Failed!*"
			Backup System Activation		B	Replace GCC with backup computer and resume monitoring.
11	Misinput During Flight (Human Error)	Unauthorized control input	GSO	Remote Control Permissions	A	Restrict RC access during the show.
					B	Monitor for irregularities.

(*Continued*)

TABLE 6.2 (*Continued*)
Examples of Response during UAS Contingency Procedures

#	Contingency Condition	Possible Failure	Crew	Item	#	Action
12	Navigation Failure	Cannot locate waypoint	GSO	Waypoint Error	A	Automatically return drone to home.
13	Position or Altitude Outside Geofence	Breach of geofence limits	GSO	Boundary Monitoring	A	Drone automatically returns to home.
14	QR Code Not Visible	Poor visibility for QR detection	GSO	Landing System	A	Use GPS positioning for safe landing.
					B	The plane will land directly on the ground
15	Rollover Protection	Tilt angle exceeds 45 degrees	GSO	Stability Monitoring	A	Drone automatically restricts tilt angle.
16	Tracking Failure	Distance exceeds parameters	GSO	Navigation Monitoring	A	Monitor ground station alerts.
					B	Return drone manually if distance is excessive.
17	Unexpected Adverse Weather Conditions	Adverse weather	GSO	Weather Assessment	A	Call Out: "*Adverse Weather!*"
			Automatic Flight		B	Takeover manual control
			Safe Landing		C	Land the aircraft
			Severe Weather		D	Initiate termination (FTS)
18	Unexpected Behavior of UAS within Geography	Flight Path Deviation	GSO		A	Call Out: "*Warning! Warning! Warning!*"
			Automatic Flight		B	Takeover manual control
			Manual Control Restored		C	Land the aircraft
			Persistent Issues		D	Initiate termination (FTS)
19	Uninvolved Persons in Controlled Ground	Safety of Area	GSO / Ground Staff		A	Call Out: "*Uninvolved Persons in the Area!*"
			Controlled Ground Breach		B	Initiate landing
			Resume Operations		C	Ensure conflict will not occur again before resuming.

AAM CONTINGENCY RESPONSE MATRIX

A Contingency Response Matrix is a 2×2 decision-support tool mapping Incident Severity against Response Capability, adapted from the International Risk Governance Council's (IRGC) RAG framework for the Respond dimension (Hollnagel, 2009, 2015). It guides stakeholders to select one of four strategies:

- Low Severity/High Capability → Routine Handling
- High Severity/High Capability → Rapid Escalation
- Low Severity/Low Capability → Monitor & Prepare
- High Severity/Low Capability → Immediate Fallback & External Support

This quadrant approach could assist organizations in allocating resources efficiently, calibrating contingency plans, and ensuring robust fallback options under critical scenarios. In the matrix, Severity is a normalized 0–1 number representing operational impact (safety risk, mission failure, public exposure).

Figure 6.3 depicts the AAM contingency response matrix, with Severity representing operational impact (safety risk, mission failure, public exposure) and the Capability computed from a weighted AAM rubric (people, procedures, detection, communications, technical, external, recovery). Three illustrative examples for incidents are also depicted in Figure 6.3 based on the following scenarios and assumptions, while considering that the risk distribution is based on people, procedures, detection, communications, technical, external, and recovery:

- Incident A: Low severity/high capability: A minor telemetry glitch during cruise was detected and cleared by the operations team; the operator has strong people, procedures, comms, and technical resilience (scores in the example reflect that), so internal routine handling is sufficient.
- Incident B: High severity/high capability: A significant avionic fault affecting navigation requiring rapid escalation to senior ops and command-and-control, but the operator has the resources and redundancy to manage it; high severity value (0.9) but moderate-to-high capability scores, so rapid internal escalation is feasible.
- Incident C: High severity/low capability: Lost-link combined with degraded DAA, and inadequate ground support due to limited trained staff, in an unfamiliar urban corridor, requiring immediate fallback and external support.

CASE STUDY: RAPID DEPLOYMENT ACROSS THE WORLD

Asia has emerged as a major test-bed for AAM, with several countries aggressively pursuing deployments of autonomous air taxis and delivery drones. These real-world implementations provide case studies in *responsive operations* and resilience engineering—i.e., how AAM systems detect and handle anomalies to maintain safety and service continuity.

TABLE 6.3
Examples of Response during UAS Emergency Procedures

#	Emergency Condition	Possible Failure	Crew	Item	#	Action
1	Drone Crash/ Collision	Collision with object or bird	Pilot	Aircraft Damage	A	Identify faulty drones.
					B	Use Pilot Escort or LAND mode for individual return.
2	Drone Operations Halt	Operator aborts due to issues	GSO	Flight Suspension	A	Switch UAS to hover mode.
					B	Return in batches using manual or automated commands.
3	GNSS Failure	GPS signal unavailable	GSO	Navigation System	A	UAS cannot continue.
					B	Initiate manual override to land or hover.
4	Gyroscope Failure	Unstable flight; potential shaking	Pilot	Stability Sensors	A	Switch to manual control.
					B	Safely land the drone.
5	Loss of Link	Communication Link Loss	GSO	Communication	A	Call Out: "*Link Lost!*"
			Signal Reconnection		B	Check signal and attempt to reconnect.
			Persistent Link Loss		C	Initiate termination (FTS)
6	Magnetometer Failure	Inaccurate heading or direction	Pilot	Heading Sensor	A	Take manual control.
					B	Land the drone safely.
7	Motor Failure	Drone loses lift	Pilot	Propulsion System	A	Take manual control.
					B	Land immediately.
8	Position or Altitude Outside Geocage	Breach of geocage limits	GSO	Boundary Monitoring	A	Drone automatically terminates its flight with the activation of the FTS.
9	Wi-Fi and GPS Jamming	Signal interference	GSO	Communication Stability	A	Detect interference.
					B	Land all drones immediately if GPS is disrupted.
10	GSO or RP Incapacitation	Health Issues	GSO or Pilot	UAS Monitoring and Control	A	Provide First Aid to the affected personnel and call an ambulance.
					B	Initiate aircraft auto return

TABLE 6.3 (*Continued*)
Examples of Response during UAS Emergency Procedures

#	Emergency Condition	Possible Failure	Crew	Item	#	Action
11	Battery fire on the ground	Power short-circuit	Ground Crew	Electrical System	A	Use the fire extinguisher to cease the fire.
12	Battery fire in the air	Power short-circuit	Pilot	Electrical System	A	Take manual control.
					B	Land the drone safely.

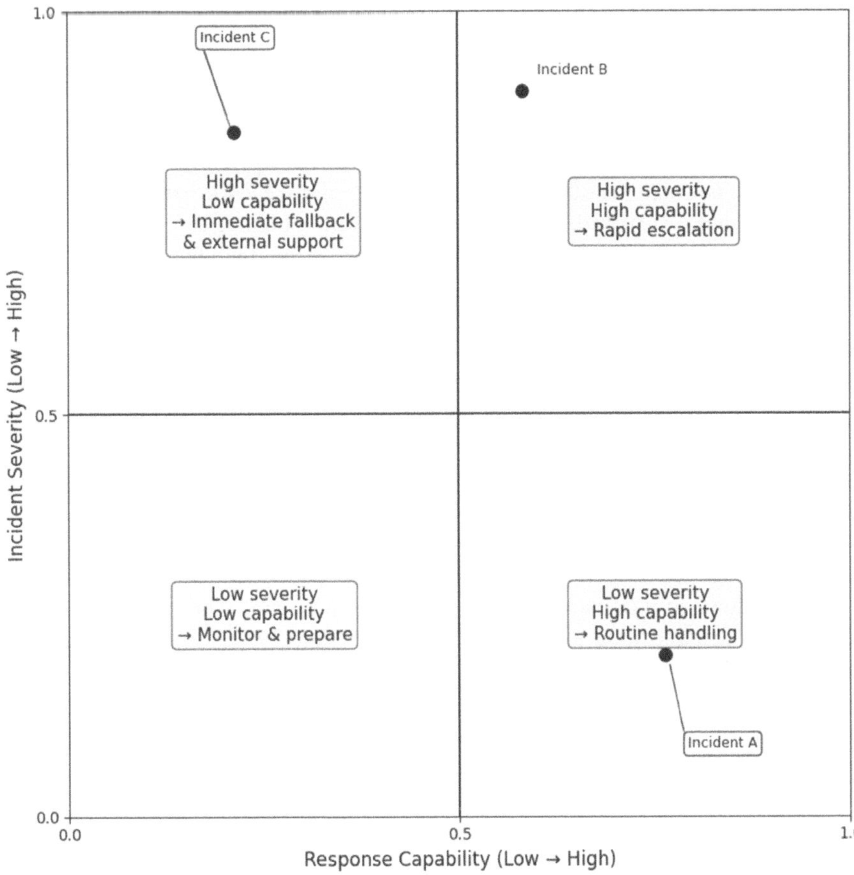

FIGURE 6.3 AAM contingency response matrix. (Credit: Author, Data source: Barrett & Upton, 2018; Florin, 2016; Hollnagel, 2015; International Risk Governance Council (IRGC), 2018; Jovanovic, 2018; Linkov et al., 2016; Petit, 2018; Seager et al., 2016; Willis & Loa, 2015; Young & Wilkinson, 2018.)

China

In particular, China's efforts stand out. Chinese companies, such as EHang (in passenger transport) and Meituan (in drone delivery), have conducted thousands of autonomous flights, refining robust response mechanisms to address in-flight issues.

EHang, the Guangzhou-based AAM pioneer, has logged tens of thousands of autonomous air taxi flights with its two-seat EH216 eVTOL (EHang, n.d.). The aircraft flies fully autonomously; all contingencies are handled either by onboard automation or by a networked ground control center that streams real-time telemetry—position, route, and system health—to operators in Guangzhou. This centralized oversight enables engineers to detect anomalies early and intervene swiftly, demonstrating high operational situational awareness.

If an abnormality occurs—such as a route deviation or component alert—controllers can (a) issue direct commands to the vehicle, (b) geofence nearby airspace to clear conflicting traffic, or (c) allow the EH216's pre-programmed failsafe modes (e.g., automatic diversion to a designated landing site) to run. The aircraft also features an autonomous emergency landing system and an onboard parachute to protect occupants in worst-case scenarios.

In collaboration with the Civil Aviation Administration of China (CAAC), EHang mapped *safe landing sites* (parks, large rooftops) along each urban route. Because these sites are pre-surveyed, the flight computer does not have to "hunt" for a location when trouble strikes; it already knows where to go, thereby cutting response time and risk (EHang, 2020).

The program's maturity was demonstrated when the EH216-S became the first fully autonomous eVTOL to earn CAAC type certification in October 2023 (EHang, 2023a). After more than 40,000 test flights, CAAC issued China's first unmanned passenger-drone operator certificate to EHang in March 2025, clearing the way for paid public flights in Guangzhou and Hefei (*Xinhua*, 2025). To support early commercial services, EHang is deploying additional route surveillance and on-site observers during the initial rollout phase (EHang Holdings Limited, 2023b). Together, advanced automation, human oversight, and engineered contingencies form a multilayered *Respond* capability.

China's low-altitude ecosystem also includes Meituan, whose drone network supports on-demand food and parcel delivery. By late 2024, Meituan had opened 53 BVLOS delivery routes in Beijing, Shenzhen, Shanghai, and Guangzhou and surpassed 200,000 commercial deliveries—nearly double the 2023 total (Wang, 2025). In the same year, its overseas affiliate, Keeta, won a Dubai BVLOS license, demonstrating the exportability of the concept (Wang, 2025).

Meituan operates a centralized drone control center that integrates air traffic data, dispatch algorithms, and safety monitoring. Each UAV incorporates a *three-tier safety architecture*, comprising redundant critical components, an automated safe-forced landing routine that diverts the drone to the nearest verified site, and a whole-aircraft parachute in the event of power or stability loss (Yangpu District Government, 2024). These fail-safes trigger automatically, while human operators supervise and coordinate with local authorities as needed.

Operational data indicate that the drones remain airworthy in moderate rain, snow, and Beaufort Force 6 winds, producing only ~51 dB of noise at a 50 m distance—a factor crucial for public acceptance (Yangpu District Government, 2024). The system's resilience was further demonstrated in August 2024, when Meituan drones delivered emergency food and medical kits to hikers at the Badaling section of the Great Wall, an area with limited road access (Yang, 2024).

Regulatory confidence has kept a similar pace of progressive implementation: in April 2025, the CAAC approved nationwide low-altitude drone delivery operations for Meituan, citing its robust operational history across 53 routes and over 450,000 deliveries (Cheng, 2025).

United Arab Emirates

In the Middle East, Dubai emerged as an early adopter of innovative urban air mobility technologies. Dubai's Roads and Transport Authority (RTA), in partnership with German company Volocopter, began trials of an autonomous air taxi (AAT) service in 2017 (Ong, 2017). Dubai has set a multi-year roadmap aimed at integrating AATs into its transit network, with a goal of achieving 25% of all trips being autonomous by 2030. A crucial part of this plan was developing policies and regulations for AAT operations during the trial period, with a strong emphasis on safety management. The RTA enlisted U.S.-based aviation safety firm JDA Aviation to oversee trial preparations and manage safety, ensuring compliance with international aviation safety standards (Volocopter, 2017).

Through these trials, Dubai implemented rigorous safety procedures. For example, each test flight of the Volocopter AAT was accompanied by a crewed chase helicopter for observation and potential intervention. The RTA also imposed strict weather limits—flights would automatically abort or be postponed if winds or other conditions exceeded safe thresholds (a critical consideration given Dubai's extreme climate) (Volocopter, 2017). The Volocopter vehicle itself was equipped with multiple redundant safety features, including nine independent battery packs and an emergency parachute system. This high degree of redundancy meant that if one or even several battery units failed, the craft could still rely on the remaining units to stay aloft. Likewise, the built-in parachute could deploy in case of a severe malfunction, allowing the vehicle to descend more safely. These measures effectively mitigated many technical risks through *redundancy* and *fail-safes*. Dubai authorities also addressed public safety and communication: because the air taxis might fly at low altitude over populated areas, the RTA developed public guidelines on what to do if an unmanned air taxi landed unexpectedly. By educating citizens, for instance, advising them to keep a safe distance if they see a drone making an emergency landing (since rotors could still be spinning), the community became part of an extended safety net during the trials. This community engagement helped avoid further accidents and built public trust in the new technology. Notably, Dubai's proactive approach has continued into the present; as of 2025, the city remains at the forefront of urban air mobility. It has even partnered with Joby Aviation to launch what could be the world's first commercial air taxi service by 2026 (Roads and Transport

Authority, 2024), building on the foundation of safety and policy frameworks established during the Volocopter trials.

Africa: Rwanda and Ghana's Experience

In Africa, AAM has primarily taken the form of drone delivery services, with notable success in countries such as Rwanda and Ghana. Companies such as Zipline International have partnered with health ministries to deliver medical supplies (blood products, vaccines, medications) to remote clinics, integrating these drone operations into national health supply chains (Gavi, 2016; Zipline, n.d.-a). These programs have been running for several years and have collectively logged thousands of deliveries, significantly reducing transport times for lifesaving supplies. A key aspect of their safety and reliability is the way they handle in-flight failures or emergencies. Zipline's fixed-wing delivery drones are designed to avoid densely populated areas whenever possible, and they are equipped with *parachute recovery systems* as a last resort. If a drone encounters a critical failure, it will deploy a parachute and glide down to the ground rather than free-falling. Importantly, the drones' flight paths are planned such that an emergency landing is likely to occur over open fields or sparsely populated areas, minimizing risk to people on the ground (Mutisya & Oloko, 2022).

Community involvement and response form another pillar of these operations' safety strategy. In Rwanda, it is an established practice that if a Zipline drone must deploy its parachute and descend outside its intended landing zone, residents assist in the recovery. Each Zipline drone has instructions and a phone number printed on it, advising bystanders to stay back and not touch the aircraft, and providing a hotline to call Zipline's response team for immediate assistance (City of Anna, 2026; Zipline, n.d.-b). Upon such a call, a trained recovery crew is dispatched swiftly to retrieve the drone and its cargo. (In many cases, community members have even carried a downed drone to the nearest health facility, treating it like "*an ambulance in the sky.*") This low-tech but effective human response mechanism leverages community awareness and goodwill. By educating locals about the drones' purpose (delivering critical medical supplies) and the proper steps to take if a drone makes an unscheduled landing, Zipline turns the public into partners in the operation. The result has been an excellent safety record for these programs—Rwanda's Zipline service has operated for years without serious injury to the public, demonstrating that robust yet straightforward response measures (such as parachutes, careful route planning, and community engagement) can enable safe drone operations even in areas with limited infrastructure. The success in Rwanda and Ghana has become a model for other regions: as of 2024, Zipline and similar services have expanded to other African countries and even to the United States, continuing to refine these safety and emergency response practices (Zipline, 2023).

Australia

In Australia, Alphabet's Wing (a drone delivery subsidiary of Google) has been delivering packages via drone to suburban areas, notably in Canberra (Australian Capital Territory) and Logan (Queensland), and has moved from trial phase to commercial service in recent years (Civil Aviation Safety Authority [CASA], 2024; Wiggers, 2022; Wing Team, 2019). Wing coordinates closely with Australia's Civil Aviation Safety Authority (CASA) and local authorities to ensure safety and compliance in these operations. For instance, Wing's drones fly predetermined routes and have defined contingency procedures. If a drone needs to abort a delivery mid-mission, there are contingency routes and designated "rally points" (predefined safe locations) where the drone can loiter or land safely away from people and property. These contingency plans are part of the operational design to handle emergencies like communication loss or system faults—rather than flying unpredictably, the drone will navigate to a *safe holding point* or *return to base* if something goes wrong. So far, Wing's operations in Australia have had few reportable incidents, aside from a well-publicized case of a drone crash into power lines in 2021 (which led to temporary power loss in part of Logan). CASA imposed strict safety conditions on Wing's service, including limits on operating hours and requirements for technical reliability, before granting approvals (CASA, 2021).

One interesting aspect Wing developed is an automated system for handling noise complaints and community feedback. While not an emergency safety response per se, it is a real-time response to community concerns, which is crucial for the sustainability of the service. Drone noise became the single most significant source of negative feedback during Wing's trials—residents in some areas complained that frequent drone overflights sounded like "mosquitoes" or a "dentist's drill" (Stonor, 2022). In response, Wing introduced a quieter drone model in 2021, which reduced perceived noise levels by approximately half. Moreover, Wing has set up channels for the public to easily lodge noise complaints (via an app or website), and it uses these inputs to adjust its operations. For example, if a particular neighborhood registers numerous noise complaints, Wing can alter flight altitudes, routes, or timing in that area to mitigate disturbance. They might fly higher above sensitive communities or avoid early morning deliveries in those neighborhoods. This responsive approach—essentially maintaining a social license to operate—has been effective: Australian regulators noted that the number of complaints during extended operations has remained relatively low, and many residents have come to appreciate the service's convenience. By addressing public concerns in near-real time, Wing reduces the risk of community backlash. In a broad sense, maintaining public trust and acceptance is part of a *resilient operation*. Even if the drones perform flawlessly technically, a service can be disrupted by public opposition. Wing's experience shows that rapid *response* to community feedback (such as adjusting flight paths for noise reduction) is integral to the success and safety of drone delivery programs.

Across these global examples, a common theme is *collaboration* and *preparedness*. Whether it is an operator working with city authorities or a regulator setting conditions for operation, everyone acknowledges that things can go wrong and that a *rapid*, *coordinated* response is the key to containing any adverse effects. It is also evident that technology alone is not a silver bullet—the human and organizational arrangements (the social side of resilience) are just as vital. AAM services that have succeeded in scaling without major incidents have done so because they prepared thoroughly and involved the right stakeholders: engineers built reliable machines, operators trained their staff and had backup plans, regulators provided oversight and flexibility for innovative mitigations, and communities were informed and became part of the shared vision of exploiting the capabilities that AAM has to offer.

REFERENCES

Ackerman, E. (2021, October 21). Air taxis are safe—According to the manufacturers. *IEEE Spectrum.* https://spectrum.ieee.org/air-taxis-are-safe-according-to-the-manufacturers

Ackerman, E., & Koziol, M. (2019, April 30). In the air with Zipline's medical delivery drones. *IEEE Spectrum, 56*(5), 24–31. https://spectrum.ieee.org/in-the-air-with-ziplines-medical-delivery-drones

Adriaensen, A., Patriarca, R., Smoker, A., & Bergström, J. (2019). A socio-technical analysis of functional properties in a joint cognitive system: A case study in an aircraft cockpit. *Ergonomics, 62*(12), 1598–1616. https://doi.org/10.1080/00140139.2019.1661527

Airbus. (2023, February). *Safety innovation: Flight envelope protection.*

Azadeh, A., Salehi, V., Ashjari, B., & Saberi, M. (2014). Performance evaluation of integrated resilience engineering factors by data envelopment analysis: The case of a petrochemical plant. *Process Safety and Environmental Protection, 92*(3), 231–241. https://doi.org/10.1016/j.psep.2013.03.002

Azadeh, A., Asadzadeh, S. M., & Tanhaeean, M. (2017). A consensus-based AHP for improved assessment of resilience engineering in maintenance organizations. *Journal of Loss Prevention in the Process Industries, 47,* 151–160. https://doi.org/10.1016/j.jlp.2017.02.028

Bagheri, S., Theunissen, E., & Curran, R. (2021). Urban air mobility safety risk assessment using Monte Carlo simulation. *Journal of Air Transport Management, 96,* 102100. https://doi.org/10.1016/j.jairtraman.2021.102100

Barrett, C. B., & Upton, J. (2018). Resilience as an integrative, measurable concept for humanitarian and development programming. In IRGC Resource Guide on Resilience (Vol. 2). Lausanne, CH: EPFL International Risk Governance Center. https://barrett.dyson.cornell.edu/files/papers/2018%20IRGC%20Resilience%20Guide%20Vol2%20%28Barrett%20%26%20Upton%29.pdf

Cheng, Y. (2025, April 22). *Meituan secures first nationwide low-altitude logistics operating license.* China Daily.

City of Anna. (2025, December 01). *Zipline drone delivery coming to Anna* (Newsflash). https://www.annatexas.gov/m/newsflash/home/detail/978

Civil Aviation Safety Authority. (2021). *Drone delivery services.* https://www.casa.gov.au/drones/industry-initiatives/drone-delivery-services

Civil Aviation Safety Authority. (2024, October 11). *Drone delivery services.* https://www.casa.gov.au/drones/industry-initiatives/drone-delivery-services

Cohen, N. (2013, November 23). Electric two-seater Volocopter is tested in Germany. *Phys.org.* https://phys.org/news/2013-11-electric-two-seater-volocopter-germany-video.html

Creedy, S. (2022, February 4). The lessons of history: How resilience distilled from experience saved QF32. *Flight Safety Australia.* https://www.flightsafetyaustralia.com/2022/02/the-lessons-of-history-how-resilience-distilled-from-experience-saved-qf32/

Dehais, F., Causse, M., Vachon, F., Régis, N., Menant, E., & Tremblay, S. (2014). Failure to detect critical auditory alerts in the cockpit: Evidence for inattentional deafness. *Human Factors, 56*(4), 631–644. https://doi.org/10.1177/0018720813510735

Edwards, T., Verma, S., & Keeler, J. (2019). Exploring human factors issues for urban air mobility operations. In *AIAA Aviation 2019 Forum* (Paper 2019–3629). American Institute of Aeronautics and Astronautics. https://ntrs.nasa.gov/api/citations/20190030706/downloads/20190030706.pdf

Edworthy, J., Stanton, N., & Hellier, E. (1995). Warnings in research and practice. *Ergonomics, 38*(11), 2145–2154. https://doi.org/10.1080/00140139508925258

EHang. (2020, May 27). EHang obtained the world's first commercial pilot operation approval of passenger-grade AAVs for air logistics uses [Press release]. *EHang.* https://www.ehang.com/news/649.html

EHang Holdings Limited. (2023a, October 13). *EHang successfully obtains type certificate for EH216-S passenger-carrying UAV system issued by Civil Aviation Administration of China* [Press release].

EHang Holdings Limited. (2023b, October 30). *EHang continues to promote operations and commercial deployment of EH216-S passenger-carrying unmanned aircraft system.* https://www.ehang.com/news/1009.html

European Union Aviation Safety Agency. (2019). *Special condition for small-category VTOL aircraft (SC-VTOL) (Issue 1).* Author. https://www.easa.europa.eu/sites/default/files/dfu/SC-VTOL-01.pdf

European Union Aviation Safety Agency. (2021, May 12). *Means of compliance with the Special Condition VTOL* (Doc. No. MOC SC-VTOL, Issue 2).

Federal Aviation Administration. (2011). *Flightcrew alerting (Advisory Circular AC 25.1322-1).* Federal Aviation Administration. https://www.faa.gov/documentLibrary/media/Advisory_Circular/AC_25.1322-1.pdf

Federal Aviation Administration. (2022). *Part 107 waiver: Zipline International Inc. (No. 107W-2022-02297).* Federal Aviation Administration. https://www.faa.gov/media/30186

Florin, M.-V. (2016). Resilience in IRGC's recommendations for risk governance. Lausanne, CH: International Risk Governance Council (IRGC). https://irgc.org/wp-content/uploads/2018/09/Florin-Resilience-in-IRGCs-Recommendations-for-Risk-Governance.pdf

Fowler, J. G., Fernandes, D. C., & Cardoso-Júnior, M. M. (2024). FRAM application for analysis of the concept of eVTOL operation in urban centers: Challenges and risks. In *XXI SITRAER (Air Transportation Symposium).* https://doi.org/10.29327/xxi-sitraer-2024.883625

Fraher, A. L. (2011). Hero making as a defense against the anxiety of responsibility and risk: A case study of US Airways Flight 1549. *Organisational & Social Dynamics, 11*(1), 59–78. https://www.ingentaconnect.com/content/phoenix/osd/2011/00000011/00000001/art00005

Functional Resonance Analysis Method. (n.d.). *Welcome to FRAM.* Retrieved January 24, 2026, from https://functionalresonance.com/

Gavi, the Vaccine Alliance. (2016, October 13). *Rwanda launches world's first national drone delivery service powered by Zipline.* https://www.gavi.org/news/media-room/rwanda-launches-worlds-first-national-drone-delivery-service-powered-zipline

Hollnagel, E. (2009). The four cornerstones of resilience engineering. In C. P. Nemeth, E. Hollnagel, & S. W. A. Dekker (Eds.), *Resilience engineering perspectives, Volume 2: Preparation and restoration* (pp. 117–133). Ashgate.

Hollnagel, E. (2011). *Resilience engineering in practice: A guidebook.* Ashgate.

Hollnagel, E. (2012). *FRAM: The functional resonance analysis method—Modelling complex socio-technical systems.* Ashgate.

Hollnagel, E. (2015a). *Introduction to the Resilience Analysis Grid (RAG)* (Technical note).

Hollnagel, E. (2015b). The resilience analysis grid (RAG): Outline (Version 2). *ErikHollnagel.com.* https://erikhollnagel.com/ideas/the-resilience-analysis-grid.html

Hollnagel, E., Woods, D. D., & Leveson, N. (Eds.). (2006). *Resilience engineering: Concepts and precepts.* Ashgate.

International Civil Aviation Organization. (2018). *Safety management manual (SMM) (doc 9859*, 4th ed.). International Civil Aviation Organization. https://store.icao.int/en/safety-management-manual-smm-doc-9859

International Risk Governance Council (IRGC). (2018). *IRGC resource guide on resilience, volume 2: Domains of resilience for complex interconnected systems*. EPFL International Risk Governance Center. https://irgc.org/wp-content/uploads/2019/01/IRGC-2018.-Resource-guide-on-resilience-vol.-2-Domains-of-resilience-for-complex-interconnected-systems.pdf

Jovanovic, A. (2018). Use of indicators for assessing resilience of critical infrastructures. In IRGC Resource Guide on Resilience. Lausanne, CH: EPFL International Risk Governance Center. https://irgc.org/wp-content/uploads/2018/09/Jovanovic-Use-of-Indicators-for-Assessing-Resilience-of-Critical-Infrastructures.pdf

Leveson, N. G. (2012). *Engineering a safer world: Systems thinking applied to safety*. MIT Press. https://direct.mit.edu/books/oa-monograph/2908/Engineering-a-Safer-WorldSystems-Thinking-Applied

Linkov, I., Trump, B. D., & Fox-Lent, C. (2016). Resilience: Approaches to risk analysis and governance. In IRGC Resource Guide on Resilience (Vol. 1). Lausanne, CH: EPFL International Risk Governance Center. https://irgc.org/wp-content/uploads/2018/09/Linkov-Trump-Fox-Lent-Resilience-Approaches-to-Risk-Analysis-and-Governance.pdf

Mutisya, L., & Oloko, S. (2022). Community engagement for drone safety in Rwanda. In *What's ethics got to do with it?* Endeva. https://endeva.org/publication/whats-ethics-got-to-do-with-it

National Academies of Sciences, Engineering, and Medicine. (2020). *Advancing aerial mobility: A national blueprint*. The National Academies Press. https://doi.org/10.17226/25646

National Aeronautics and Space Administration, Ames Research Center. (2020). *Urban Air Mobility Grand Challenge: Human-in-the-Loop simulation overview (NASA TM 20205008970)*. National Aeronautics and Space Administration. https://ntrs.nasa.gov/citations/20205008970

Nemeth, C. P., & Holbrook, J. (2021). Resilience engineering's potential for advanced air mobility (AAM). In *Proceedings of the 72nd international symposium on aviation psychology (ISAP 2021)* (pp. 122–127). NASA. https://ntrs.nasa.gov/citations/20210012631

Ong, T. (2017, September 26). Dubai starts testing crewless two-person "flying taxis". *The Verge*. https://www.theverge.com/2017/9/26/16365614/dubai-testing-uncrewed-two-person-flying-taxis-volocopter

Pariès, J. (2009). *Lessons from (the) Hudson. HindSight* (Issue 9, Summer), 26–28.

Patriarca, R., Di Gravio, G., Woltjer, R., Costantino, F., Praetorius, G., Ferreira, P., & Hollnagel, E. (2020). Framing the FRAM: A literature review on the functional resonance analysis method. *Safety Science*, *129*, 104827. https://doi.org/10.1016/j.ssci.2020.104827

Petit, O. (2018). Resilience assessment in homeland security. In IRGC Resource Guide on Resilience (Vol. 2). Lausanne, CH: EPFL International Risk Governance Center. https://irgc.org/wp-content/uploads/2018/12/Petit-for-IRGC-Resilience-Guide-Vol-2-2018.pdf

Roads and Transport Authority. (2024, March). *Al Masar* (Issue No. 189). Roads and Transport Authority. https://www.rta.ae/links/magazine/masar/Al_Masar_189_Eng.pdf

Sarter, N. B., & Woods, D. D. (1995). How in the world did we get into that mode? Mode error and awareness in supervisory control. *Human Factors*, *37*(1), 5–19. https://doi.org/10.1518/001872095779049516

Seager, T. P., Clark, S. S., & Eisenberg, D. A. (2016). A multidimensional review of resilience: Resources, processes, and outcome priorities. In IRGC Resource Guide on Resilience (Vol. 1). Lausanne, CH: EPFL International Risk Governance Center. https://irgc.org/wp-content/uploads/2018/09/Seager-et-al.-A-Multidimensional-Review-of-Resilience.pdf

Stonor, C. (2022, September 28). Wing drone delivery noise: Australians complain "It's gone too far!". *eVTOL Insights.* https://evtolinsights.com/2022/09/wing-drone-delivery-noise-australians-complain-its-gone-too-far

Thomas, J. P., & Leveson, N. G. (2013). *Safety analysis of complex systems: A practical STPA approach* (MIT Technical Report ESD-2013-02). Massachusetts Institute of Technology. https://psas.scripts.mit.edu/home/get_file.php?name=STPA_handbook.pdf

Volocopter. (2017, September 25). First ever public demonstration of an autonomous urban air taxi—German startup Volocopter writes aviation history in Dubai [Press release]. *Volocopter Press Center.* https://www.volocopter.com/cn/newsroom/first-ever-public-demonstration-of-an-autonomous-urban-air-taxi

Wang, Y. (2025, February 4). Delivery drones: Meituan and competitors expand presence. *Dao Insights.* https://daoinsights.com/news/delivery-drones-meituan-and-competitors-expand-presence/

Wiggers, K. (2022, November 8). *Wing brings drone delivery options to DoorDash customers in Logan, Australia. TechCrunch.* https://techcrunch.com/2022/11/08/wing-brings-drone-delivery-options-to-doordash-customers-in-logan-australia/

Willis, H. H., & Loa, K. (2015). Measuring the resilience of energy distribution systems (RR-883-DOE). Santa Monica, CA: RAND Corporation. https://www.rand.org/content/dam/rand/pubs/research_reports/RR800/RR883/RAND_RR883.pdf

Wing, D. J., Chancey, E. T., Politowicz, M. S., & Ballin, M. G. (2020). *Achieving resilient in-flight performance for advanced air mobility through simplified vehicle operations.* AIAA AVIATION Forum (virtual event). NASA Langley Research Center. NASA Technical Reports Server.

Wing Team. (2019, April 8). *Wing launches commercial air delivery service in Canberra.* Wing. https://wing.com/news/wing-launches-commercial-air-delivery

Wing Team. (2024, February 28). *Wing's multi-layered approach to detect & avoid.* Wing.

Xinhua. (2025, March 30). China issues 1st batch of passenger drone operation certificates. *Xinhua.* https://english.news.cn/20250330/3d1933a3872e40d8a5f4d0bfbe912992/c.html

Yang, Y. (2024, December 23). Highlights of new quality productive forces: Low altitude economy. *China Daily.* https://enapp.chinadaily.com.cn/a/202412/23/AP6768b67aa3103f0966a0d2df.html

Yangpu District Government. (2024, July 26). People's City fifth anniversary: Low altitude economy "takes off," Meituan drone Huangxing Park route enters trial [Government news release]. *Yangpu District Government.* https://english.shyp.gov.cn/ywb/lntc5zn/20240808/461195.html

Young, M., & Wilkinson, A. (2018). Managing energy transition through dynamic resilience. In IRGC Resource Guide on Resilience (Vol. 2). Lausanne, CH: EPFL International Risk Governance Center. https://irgc.org/wp-content/uploads/2018/12/Young-et-al-for-IRGC-Resilience-Guide-Vol-2-2018.pdf

Zipline. (n.d.). Zipline expands deliveries in Rwanda to boost economic development & wildlife conservation. Zipline Newsroom. Retrieved October 25, 2025, from https://www.zipline.com/newsroom/news/announcements/zipline-expands-deliveries-in-rwanda-to-boost-economic-development-wildlife-conservation

Zipline. (n.d.-a). *How Zipline is bringing Ghana's remote communities within reach.* https://www.zipline.com/newsroom/stories/articles/zipline-ghana-remote-communities-within-reach

Zipline. (n.d.-b). *Response team.* Retrieved January 24, 2026, from https://www.zipline.com/responseteam

7 Learn
Continuous Growth

INTRODUCTION

At 07:53 LT on a fog-covered winter morning, a small quad-rotor loaded with urgent blood supplies lost GPS reception while threading the "urban canyon" between two downtown hospitals. Instead of drifting into a busy arterial road, the aircraft executed a pre-programmed hover, the remote pilot immediately switched to visual orienting cues, and the mission software uploaded a contingency route that steered clear of surrounding high-rise cranes. The episode ended without injury or property damage, but the voluntary report the pilot filed that afternoon triggered something more significant: a fleet-wide firmware patch and a new checklist item requiring pilots to verify alternative navigation modes before launch (NASA ASRS, 2024).

This quiet, near-miss illustrates the central premise of resilient safety in advanced air mobility (AAM): learning is not confined to spectacular failures; it thrives on the dozens of *almost-events* that happen every week (see Figure 7.1). Traditional Safety-I thinking aims for the *absence* of accidents; Safety-II aims for the *abundance* of operations that go right by understanding how everyday adaptations succeed (Hollnagel, 2014). Turning a single drone's GPS hiccup into a system-wide improvement demands robust mechanisms that capture front-line experience and convert it into collective knowledge.

Tools such as the Functional Resonance Analysis Method (FRAM) help analysts model how normal variations—wind shear over a river bend, latency in a 5G link, pilot workload during shift change—can resonate into unwanted outcomes. Viewed through a FRAM lens, the morning's incident becomes a case study in *work-as-done*: how humans and machines co-adapt under real-world complexity, and how that adaptation can be engineered into safer standard practice.

LEARNING FROM INCIDENTS: MICRO-LEVEL INSIGHTS IN AAM

One of the richest sources of learning in any safety-critical domain is operational incidents—the minor errors, surprises, and recoveries that occur in day-to-day practice. In nascent AAM operations (including unmanned aircraft systems), pilots and operators are already encountering unplanned situations and documenting them. For example, NASA's Aviation Safety Reporting System (ASRS) has compiled 50 UAS incident reports to share lessons learned with the community (NASA ASRS, 2025)—see Table 7.1 for a summary of the Key UAS Safety Trends, Causal Factors, and Mitigation Strategies, while Annex B lists substantial worldwide UAS incidents since 2016. An encouraging sign is that many UAS pilots are voluntarily reporting their mistakes and near-misses, which suggests a promising learning culture,

DOI: 10.1201/9781003546306-9

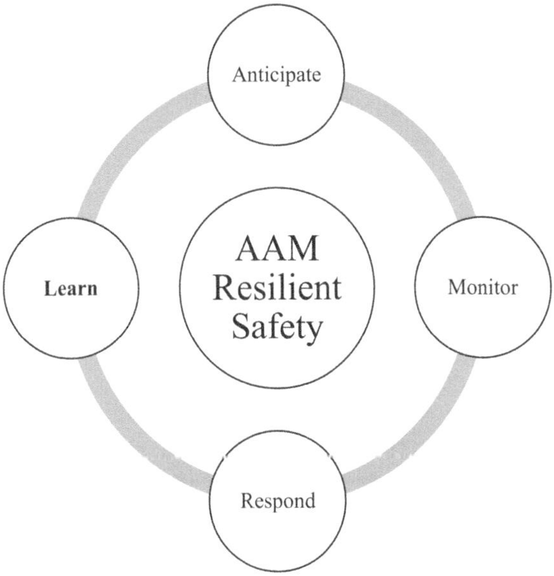

FIGURE 7.1 Learn function for resilient safety in AAM.

Note: Based on Hollnagel's (2009, 2015) four resilience abilities (anticipate, monitor, respond, learn).

as pilots are willing to share errors and lessons. Indeed, each report is seen as an opportunity to ask, *"What would I have done?"* and exercise decision-making skills (NASA ASRS, 2025). Many reports describe how the pilot changed their behavior afterward—for instance, by double-checking altitude settings or adding extra buffer time to missions. Such self-corrective actions indicate individual learning: the operator recognizes a lapse (like misreading an altitude or forgetting to obtain airspace authorization) and commits to a new practice to prevent recurrence. In one case, a drone pilot admitted they became "*too wrapped up*" in other tasks and *"neglected to file the flight [authorization]... Going forward, I will be using a checklist"* (NASA ASRS, 2025). This kind of *micro-level learning*—an individual reflecting on an incident and adjusting their routine—is the foundation of resilient performance.

Another incident that highlights the need for sufficient coordination within shared airspace by UAS and aircraft happened on February 6, 2020, when a small UAS collided with a helicopter during the videography of an off-road truck race near Johnson Valley, CA (see Figure 7.2). As per NTSB (2022), contributory factors were the sUAS remote pilot's failure to assess and mitigate the risks of operations in close proximity to another aircraft, as well as the lack of inclusion of the sUAS operations as part of the overall aviation activity and risk mitigation. As suggested by the drone operator at his post-incident report, to ensure deconfliction there is a need for: shared operations maps, clear vertical deconfliction between aircraft and UAS (aircraft $\geq$500 feet and UAS < 400 feet AGL), no UAS-zones to be briefed, pre-flight coordination briefs and continuous communication between all players, and assignment of a dedicated

FIGURE 7.2 Video snapshot before midair collision between a drone and a helicopter. (*Source:* NTSB, 2022.)

spotter (who is adding an extra layer of safety with their supplementary situational awareness) for each drone pilot to mitigate the restrictions due to low level flying and ambient noise, which can mask an approaching aircraft. Such incidents could be insightful for the emerging AAM industry to realize that solid hazard identification and risk mitigation in constrained airspace are paramount for safe and effective operations.

Yet, the incident reports also highlight concerning patterns that suggest a lack of broader organizational learning. Certain violations or errors have recurred across multiple reports, such as confusion with airspace authorizations or inadvertently flying into Temporary Flight Restrictions (TFRs) (Federal Aviation Administration [FAA], 2025a, b; NASA ASRS, 2025). The repetition of similar issues suggests that lessons learned by one operator were not systematically propagated to others. For example, one recreational UAS pilot reported accidentally flying near a stadium event TFR because they hadn't checked the app or NOTAMs beforehand, and only afterward realized their mistake and the rules they missed. Another report described a pilot who *assumed* their airspace request was approved when, in fact, it had been rejected in the app. Being out of practice with the software, they didn't catch the rejection until later and then resolved to use the app's features more carefully in the future. These kinds of repeated lapses—entering controlled airspace without clearance or missing a TFR—imply that while individual operators might be learning from their own mistakes, the wider community or companies involved are not yet fully institutionalizing those lessons (e.g., via improved training, standard procedures, or software interface fixes). This gap between individual and organizational learning is critical to address as AAM scales up.

To bridge this gap, the AAM industry must adopt structures that capture micro-level data and transform it into macro-level improvements. Traditional aviation provides a model: for decades, systems like ASRS and Flight Operations Quality Assurance (FOQA) have collected incident reports and flight data from individuals, then analyzed them for industry-wide trends (FAA Safety Team & NASA, 2016). The same is beginning in the UAS/AAM domain. NASA's ASRS is receiving UAS-related incident reports, and each submission is a learning opportunity that can be shared broadly. For instance, if one drone pilot encounters a software glitch that led to a loss of control, that report—once de-identified—can alert other operators and manufacturers to the hazard. Learning systems in AAM should encourage this kind of data sharing. In practice, that means not only collecting reports but ensuring they are reviewed and communicated. A robust process might include an internal safety review board that meets regularly to discuss recent incidents, identify root causes and common factors, and implement corrective actions across the operation. Crucially, feedback must loop back to front-line personnel. When a pilot reports an issue, they should hear about the resolution or changes that resulted. Such feedback reinforces the value of reporting and completes the learning cycle from the field to management and back (NASA ASRS, 2025).

From these *micro-level* insights, two key points emerge. First, AAM operations will experience near-misses and small failures—this is inevitable in a complex, evolving environment. But if those experiences are openly reported and analyzed, they become fuel for learning rather than precursors to disaster. Second, individual learning, while valuable, is not enough; organizations need to actively cultivate mechanisms to aggregate and disseminate lessons.

TABLE 7.1
Key UAS Safety Trends, Causal Factors, and Mitigation Strategies

Trend	Brief Description	Common Causal Factors	Mitigation Strategies
Airspace/TFR Violations	Pilots inadvertently entered restricted airspace (e.g., near airports or firefighting zones) without proper authorization.	• Inadequate preflight planning • Over-reliance on app-based *geofencing* • Confusion over TFR boundaries or LAANC time windows	• Cross-check multiple sources for TFRs/NOTAMs • Verify time/date in LAANC or FAA DroneZone before launch • Incorporate robust preflight checklists
Near Midair Collisions (NMAC)	Air carrier or GA pilots reported drones in dangerously close proximity during takeoff, climb, or approach, often triggering TCAS advisories.	• Unaware or noncompliant operators • Flight near airports without authorization • Loss of situational awareness	• Maintain strict compliance with controlled airspace requirements • Use conspicuity lighting or strobes on the drone • Coordinate closely with ATC where applicable

(Continued)

TABLE 7.1 (*Continued*)
Key UAS Safety Trends, Causal Factors, and Mitigation Strategies

Trend	Brief Description	Common Causal Factors	Mitigation Strategies
Loss of Positive Control / "Fly-Aways"	Uncommanded drone flight due to RF interference, GPS glitches, software updates, or battery malfunctions.	• Firmware/software issues • Weak or lost link (RF interference) • Inadequate failsafe programming	• Conduct thorough preflight tests, including firmware checks • Set reliable RTH altitudes and check battery health • Land immediately at the first sign of malfunction
Exceeding Altitude or Lateral Bounds	Operators mistakenly flew higher than the approved ceiling or drifted beyond their authorized radius (e.g., 50 ft LAANC grid vs. 65 ft actual).	• Terrain elevation miscalculations (AGL vs. MSL) • Inaccurate or confusing flight app settings • Pilot fatigue or haste	• Configure app settings carefully (e.g., confirm AGL vs. MSL) • Monitor altitude readouts frequently • Build buffer time into flights to avoid rushing
Community Tensions and External Interference	Neighbors or bystanders fired guns at a drone or complained angrily about perceived privacy violations; one pilot reported bullet strikes near their craft.	• Misinformation about drone overflight laws • Fear or misunderstanding of UAS operations • Low public awareness of legitimate uses	• Conduct community outreach and public education • Display clear markings/ strobes and remain visible • Avoid persistent loitering over private property unless operationally necessary
Human Factors: Stress, Fatigue, and App Reliance	Pilots continued missions while jet-lagged or under deadline pressure, leading to errors (e.g., mixing up local time, skipping final verification, or trusting flawed automation).	• Pilot exhaustion or travel-related fatigue • Organizational pressure to complete missions • Blind trust in software "safe to fly"	• Incorporate *IMSAFE* checks (Illness, Medication, Stress, Alcohol, Fatigue, Emotion) • Use multiple apps or manual cross-checks • Build rest periods and strong crew resource management into mission plans
Operating Specialized or Heavier UAS Without Waiver	Some agencies or operators flew drones >55 lbs (or performing advanced tasks) under Part 107, but did not hold the required waiver or exemption.	• Unfamiliarity with special authorization processes • Assumption that Part 107 alone suffices for all UAS • Confusion about additional waivers	• Confirm and obtain all required waivers (e.g., FAA Part 107, Part 137, or COA for public aircraft) • Ensure operational parameters match any granted exemption • Conduct internal audits to verify compliance with weight/mission-specific rules

Source: Adapted from NASA ASRS (2025) data.

HUMAN-MACHINE TEAMING (HMT) AND LEARNING IN AAM OPERATIONS

AAM will rely heavily on automation and autonomous systems, but humans remain integral to the overall system. AAM can be viewed as a joint cognitive system of humans and machines working together, which introduces the concept of human-machine teaming (HMT). Effective learning in AAM safety must therefore encompass not only human learning from incidents, but also how humans and intelligent machines learn to work together. HMT in AAM may take forms like a single pilot interacting with a highly automated eVTOL, a remote supervisor managing multiple autonomous drones, or dispatchers overseeing AI-driven flight routing systems. These configurations resemble scenarios in defense (e.g., one operator controlling a drone swarm) and manufacturing (a technician overseeing a bank of robots). In each case, team performance and safety emerge from the interaction of humans and machines, each with distinct strengths and weaknesses.

NEW ROLES AND CHALLENGES

As NASA researchers have highlighted, AAM will entail novel human roles and new kinds of human–automation interactions. For example, NASA's Langley Research Center has explored future concepts where one operator may supervise multiple highly automated vehicles or manage flights remotely. Ensuring resilience in such scenarios hinges on careful function allocation: let automation handle routine tasks and even initial failure responses, but keep humans in the loop as *adaptive problem-solvers* of last resort (Buck et al., 2025; Chancey et al., 2020).

Learning in this context has two complementary dimensions: (1) preparing the human operator—through role design and training—to supervise, maintain sufficient situation awareness, and intervene effectively when required, and (2) shaping the autonomy so its behavior is transparent and predictable, and so it can request or incorporate human guidance when conditions warrant it(Gregory & Trujillo, 2016; Neogi et al., 2025). Adjacent sectors highlight these points. In defense, where HMT is extensively studied, there is a mantra that effective HMT must be *human-centered*—the technology should be designed to support human decision-making. Humans must be trained to understand the capabilities and limitations of AI/automation. If the human operator is *out of the loop* or lacks trust in the machine, the team fails to function effectively (Gouraud et al., 2017; Neogi et al., 2025). Increasing levels of automation can unintentionally degrade human readiness by pushing operators into a passive monitoring role—an *out of the loop* condition associated with reduced attention and poorer recovery when automation behaves unexpectedly or fails. Accordingly, FAA human-factors work has highlighted knowledge/skill degradation due to lack of practice and stressed that, if pilots remain responsible for safety of flight, they must retain the knowledge and skills to intervene effectively and the system must be designed to enable such intervention (Gouraud et al., 2017; FAA, 2020).

This is directly relevant to AAM: keeping humans meaningfully engaged is not optional if they must supervise, diagnose, and take over in off-nominal conditions. Commercial aviation has already confronted a closely related challenge—highly reliable automation can contribute to complacency and erosion of manual flying proficiency—so FAA guidance and industry recommendations emphasize maintaining manual-flight competence through recurrent training, including scenario-based exercises, and operational policies that encourage appropriate manual flying practice (FAA, 2017; FAA, 2016). AAM operators, similarly, might require remote pilots to periodically take manual control in simulations to maintain their skills, ensuring that autonomy does not lull operators into complacency.

Training for Teaming and Trust

Building an effective human-machine team in AAM will require focused training and iterative learning. Pilots/operators need to develop the right level of trust in their automated teammates—neither over-trusting nor under-trusting them. Trust is calibrated through experience, as encountered in varied scenarios and observed behavior (Johnson et al., 2021; Lebiere et al., 2021). Therefore, experimentation and simulation are crucial to building trust in HMT (Endsley & Kiris, 1995). The military has found that only through repeated drills can operators learn the "breaking points" of their AI/robot teammates, understanding when the AI performs well and when it needs help (de Visser & Parasuraman, 2011; Johnson et al., 2021). AAM could adopt a similar approach: before allowing a single operator to manage, say, five autonomous air taxis, there should be extensive simulated operations where that operator experiences everyday situations and emergencies with those vehicles. Through this process, the human learns how the autonomy reacts, and the system designers learn where the automation might need to hand off control (Buck et al., 2025; Lieberman, 2022). The Atlantic Council's study on modern militaries notes that humans must develop competencies and comfort with AI systems to fully exploit their value, which only occurs when equal attention is given to training humans as is given to advancing the AI (de Visser et al., 2020; Johnson et al., 2021; Nurkin & Siegel, 2023). In practical terms, AAM companies might include dedicated HMT modules in their training—for example, teaching a remote pilot how the autonomous flight software "thinks" (its algorithms for path planning, obstacle avoidance, lost-link procedures, etc.), so the pilot can anticipate its likely actions (Artemenko, 2022). Likewise, drills where the autonomy encounters edge cases and *asks* for human guidance will teach both parties: the human learns to respond quickly and appropriately, and the AI's designers learn how to improve its request-for-help triggers (Luebbers et al., 2022; You et al., 2022).

Designing for Shared Situational Awareness

A key aspect of HMT is developing *shared situational awareness* between humans and machines. Each needs insight into the other's status and intentions. In traditional cockpits, this is addressed by good interface design (automation annunciations, mode indicators) and procedures where pilots verbally indicate the automation mode they're in ("Autopilot ON," "VNAV armed," etc.). In AAM, especially in remote or multi-vehicle operations, maintaining situational awareness is even more challenging. The operator's interface must aggregate information from multiple vehicles and highlight anomalies. If an AI is managing flight trajectories, it should convey its plan and any uncertainties to the human.

We can learn from manufacturing, where control room operators monitor dozens of processes through dashboards. The best systems use *visual analytics* and *smart alerts* to direct human attention to where it's needed, essentially teaching operators about the system's state in real-time. Similarly, in air traffic control (another domain of one-to-many human supervision), tools have been introduced that project future conflicts and alert controllers, effectively helping controllers learn about potential problems a few minutes in advance. AAM can emulate these practices. For instance, an AAM network operations center might have AI-driven prediction tools that alert the human supervisor if, for example, two flight plans might converge unsafely in five minutes, or if one drone's battery shows an anomalous discharge rate. This gives the human a chance to intervene early. Over time, the human operators also learn from these AI predictions—they start to develop a mental model of common failure patterns (e.g., "*We often get alerts about lost GPS in this downtown corridor at dusk*"). They can work with engineers to address root causes (perhaps by adding better sensors or improving map data for that area).

Multi-Robot Management

One specific HMT scenario likely in AAM is one-to-many supervision, akin to managing multiple robotic agents in other industries. In UAS operations today, experiments have shown that a single operator can oversee multiple drones in routine conditions, but workload and risk spike when anomalies occur simultaneously. To manage this, both technology and procedures need to adapt. Robotics research suggests scheduling algorithms can help stagger tasks so that not all robots demand attention at once, and robots can be endowed with more autonomy to handle minor issues independently. Meanwhile, humans must learn strategies for attention management across multiple agents. The defense sector has trialed controlling swarms of drones, finding that commanders rely on setting high-level goals and letting the autonomy handle details, intervening only by exception. The operator thus learns to function more as a mission manager than a direct pilot, which is likely how AAM remote supervisors will operate. For example, in tests of drone swarms, an operator might set waypoints or target areas, and the autonomy figures out the optimal allocation of drones to cover them; if something unexpected occurs (like one drone failing), the system alerts the human to reallocate resources. Translating this to AAM, an operator might supervise four air taxis on planned routes—the automation would normally handle spacing, sequencing for landing, and collision avoidance. However, if one vehicle encounters a severe weather cell, it would notify the operator and perhaps suggest rerouting the others. The human's role becomes managing these exceptions with the aid of AI recommendations, while continuous on-the-job learning happens as the operator experiences various combinations of issues and the AI improves its logic based on human feedback.

Cross-Industry Reflections

Both aviation and other industries emphasize that successful HMT is not a one-time setup; it is an evolving partnership. Traditional aviation's use of autopilots and flight management systems taught us that humans must understand automation logic (pilots are trained extensively on what the autopilot can and cannot do). In manufacturing, workers had to learn to safely share workspaces with collaborative robots (e.g., understanding the robot's movement patterns and trusting its safety sensors). In military domains, pilots of semi-autonomous drones learned through simulations how to hand over certain tasks to AI (like target tracking) while retaining ultimate control. All these cases reinforce that learning is *bidirectional*: humans adapt to machines, and machines (or their designers) adapt to better serve humans. AAM will require ongoing programs to study HMT performance and incorporate improvements into both training and system design. For example, early AAM operations might deliberately start with small human-to-vehicle ratios (say 1 operator: 1 vehicle or 1:2) and gradually increase as experience is gained, ensuring that we learn the limits before scaling up. During this process, data on operator workload, situational awareness lapses, and any safety incidents should be closely analyzed (perhaps using FRAM to model the human-automation interactions) to refine the teaming concept. Notably, when a UAS team experienced a near mid-air conflict with a low-flying manned aircraft despite meticulous planning, they debriefed and decided that in all high-traffic scenarios, one team member must have sole responsibility for monitoring air traffic and visually scanning for intruding aircraft. This is a prime example of *micro-level learning* in HMT: humans adjusted their teaming practice (by adding a dedicated "traffic watcher") to better complement the limitations of both automation and human situational awareness in busy airspace.

EMOTIONAL INTELLIGENCE AND MINDFULNESS IN SAFETY LEARNING

While technology and systems are critical, the human element in learning—particularly the cognitive and emotional skills of individuals—is equally important (Coronado-Maldonado & Benítez-Márquez, 2023; Geller, 2023; Gemmano et al., 2024; Stroeve et al., 2022). Two concepts gaining attention in high-reliability fields are *Emotional Intelligence (EQ)* and *Mindfulness (*Alaydi & Ng, 2024; Duplechan, 2024; Mezentseva et al., 2023; Samu et al., 2025). EQ refers to the ability to recognize and manage one's own emotions and to understand and influence the emotions of others (Coronado-Maldonado & Benítez-Márquez, 2023; Geller, 2023; Moserle et al., 2024). Mindfulness refers to a state of present-moment awareness and focus, which can be cultivated through practices like meditation or breathing exercises (Alaydi & Ng, 2024;Shonin & Van Gordon, 2014). At first glance, these may seem removed from AAM operations, but they have a direct impact on safety culture, team performance, and the capacity to learn from incidents (Alaydi & Ng, 2024; Coronado-Maldonado & Benítez-Márquez, 2023; Mezentseva et al., 2023).

EMOTIONAL INTELLIGENCE IN SAFETY TEAMS

AAM operations will involve teams, including a flight crew for a piloted eVTOL, a control room crew managing drone traffic, and maintenance and dispatch teams to ensure everything runs smoothly (Sagirli et al., 2024). High EQ among team members and leaders fosters better communication, trust, and adaptability

(Coronado-Maldonado & Benítez-Márquez, 2023; Lee et al., 2023; Boyar et al., 2023; Dugger & McCrory, 2021). In traditional aviation, crew resource management (CRM) training implicitly covers aspects of EQ: pilots are taught to speak up assertively when they perceive a problem (self-confidence) but also listen to others and encourage input (empathy, social awareness) (Clarke, 2023; Energy Institute, 2014; UK Civil Aviation Authority, 2016). These behaviors have been linked to improved safety outcomes (Robson & Manacapilli, 2014; Wang et al., 2023). Recent studies in aviation have explicitly found that EQ in leaders correlates with a positive safety culture and fewer incidents (Bates, 2023). When leaders demonstrate empathy and self-regulation, it creates an environment where frontline employees feel respected and motivated to adhere to safety practices and report concerns. In the AAM context, consider a scenario: a remote operations supervisor notices an operator is unusually quiet or tense after a near-miss incident. An emotionally intelligent supervisor might take the time to debrief supportively, acknowledging the stress and encouraging the operator to discuss what happened, rather than simply brushing it aside or focusing only on the procedure. This not only helps the individual process the event (which is vital for learning from it) but also signals to others that emotions are recognized in this workplace, reducing the stigma around admitting stress or confusion, which in turn is crucial for a healthy reporting culture (Masi et al., 2023).

Furthermore, high EQ can improve decision-making under pressure (Beattie et al., 2023; Wang et al., 2023). A pilot who can manage their fear or anger in an emergency will think more clearly and follow procedures rather than freezing or overreacting (Masi et al., 2023; Robson & Manacapilli, 2014). For example, military aviation training includes stress inoculation—building emotional resilience so that pilots remain composed when things go off-center (Gemmano et al., 2023; Robson & Manacapilli, 2014). EQ training (like identifying one's stress signals and practicing techniques to stay calm) is a part of that (Hallé et al., 2025). AAM pilots and operators may similarly benefit from training in emotional self-awareness, recognizing when workload or stress is degrading their performance and taking steps (such as communicating "unable" or requesting assistance) rather than trying to push through in a compromised state. Organizations can also emphasize EQ in their hiring and promotion for safety-critical roles. For instance, selecting supervisors who demonstrate empathy and fairness could strengthen a just culture: employees are more likely to report issues to a manager they trust to be fair-minded. Over time, a team with high EQ will likely catch and address minor safety concerns before they escalate, simply because the team communicates openly and supports each other (Bates, 2023; Wang et al., 2023).

Mindfulness for Operational Focus

Mindfulness, or present-moment focused attention, has direct implications for safety and learning. AAM operations (especially in dense urban environments or confined airspace) will be cognitively demanding. Distractions or mind-wandering could lead to missed signals or delayed reactions. Mindfulness training has been shown to improve individuals' ability to concentrate on the task at hand and maintain situational awareness. Research in other industries suggests that mindfulness practices can enhance task reliability and help workers identify subtle cues of potential danger. For example, Joyner and Lardner (2008) suggested that employing mindfulness training in industrial settings improved task reliability and allowed workers to pause and consider immediate risks before starting a job. Likewise, Glomb et al. (2011) argued that mindfulness entails receptive attention to and awareness of present events and can involve a wide attentional breadth that helps people notice a broader range of environmental stimuli in dynamic settings, with potential benefits for decision-making (Glomb et al., 2011; Dane, 2011). In essence, mindfulness helped them stay both attentive to what they were doing *and* broadly aware of their surroundings, a critical duality in aviation tasks. Additionally, full attention in the moment tends to reduce emotional reactivity—individuals are less likely to panic or become angry, and more likely to respond thoughtfully. This leads to more confident employees who are better equipped to cope with stress and uncertainty.

Applying this to AAM, for instance, a vertiport flight coordinator who has practiced mindfulness might be more adept at managing multiple incoming and outgoing vehicles without tunnel vision or overload—for instance, they might notice a subtle change in one drone's sound or behavior that signals a developing problem, while still calmly managing other routine tasks. Or a pilot facing an unexpected GPS outage may recall their mindfulness breathing techniques to steady themselves, then methodically work the backup procedure rather than getting flustered. Some airlines have begun exploring mindfulness programs for pilots and air traffic controllers to enhance focus and stress management. The oil and gas industry, for example, has piloted mindfulness training for offshore crews to improve safety, finding a reduction in incidents attributed to lapses in attention (as reported in trade publications). These sectors report that mindfulness can combat complacency and the tendency to go on "autopilot mode" during routine operations, keeping workers alert to anomalies. AAM, with its routine yet high-stakes operational tempo (particularly when autonomous systems handle many routine tasks and humans are mostly supervising), could benefit from mindfulness practices to ensure humans remain *actively engaged*—essentially, keeping the *human in the loop* mentally, even if the automation is physically in control.

Mindfulness in Organizational Learning

Beyond individual attention, mindfulness has a metaphorical role at the organizational level. In safety and resilience research, collective mindfulness (often discussed as mindful organizing) refers to an organization's shared capability to notice emerging issues and respond quickly. The concept was articulated in the HRO literature by Weick and colleagues and later popularized for practitioners by Weick and Sutcliffe, particularly through *Managing the Unexpected* (Weick et al., 1999; Weick & Sutcliffe, 2007). HRO exemplars commonly discussed in this tradition include aircraft-carrier flight decks and nuclear power control rooms/plants. *Collective mindfulness* means the organization is continually aware of the potential for failure and is preoccupied with identifying minor problems before they grow, even when operations seem to be going well. Practices that contribute to this include frequent briefings, encouraging everyone (including junior staff) to voice their concerns, and continually updating a *shared mental model* of the operation. AAM operators can cultivate this by, for example, starting each shift with a *team mindfulness briefing*—not in a spiritual sense, but rather as a pause to focus on the present operational picture. During this brief, the team could review the plan for the day, any new hazards or changes, and explicitly invite team members to speak up about anything that worries them or any "gut feelings" of unease. By legitimizing that sense of unease (what some safety experts refer to as *chronic unease* or *healthy paranoia*), leaders signal that vigilance is valued and respected. Mindfulness training can actually increase what's termed *mindful uneasiness*—meaning people remain a bit wary and thus do not take operations for granted. This prevents complacency in a rapidly evolving AAM context where new issues can arise at any time. In practical terms, an organization might integrate a 2 min ute mindfulness exercise at the start of meetings or shifts to help everyone center their attention on the safety-critical aspects of their work. Over time, these habits contribute to an organizational culture that is attentive and ready to learn from any deviation or sign of trouble.

Mindfulness Practice Implementation

How could AAM incorporate mindfulness practically? Some options include introducing short mindfulness exercises into training and daily routines. Even a 5-minute guided breathing or focusing exercise before a complex operation can help operators clear distractions and center themselves. Several studies in healthcare and the military have shown that even brief mindfulness training can improve working memory and cognitive flexibility under stress. Another approach is encouraging *micro-breaks* during operations—for example, a remote pilot supervising flights could be prompted to take a 30-second pause every hour to breathe deeply and reset their focus (air traffic controllers already have mandated break periods to manage cognitive load; during those breaks, a mindfulness technique could be suggested to maximize recovery). To ensure these "soft" practices gain traction in a very tech-driven industry, leadership endorsement is crucial. When leaders participate in and normalize these practices, it sends a message that attentiveness and well-being are valued as components of safety. An emotionally intelligent leadership team might openly discuss not only technical issues in safety meetings, but also human performance issues such as fatigue, stress, or distraction, and encourage the sharing of coping strategies. This creates psychological safety—employees know they can admit to feeling overwhelmed without fear of repercussions. They will be supported, which in turn means potential safety risks (like fatigue or overload) are more likely to be surfaced before they lead to errors.

Measuring the impact of EQ and mindfulness initiatives can be challenging, but organizations can look at indicators such as reporting rates (does reporting of near-misses improve after just culture and EQ-focused leadership training?), attention-related error rates (do we see fewer incidents of "forgot to do X" or "didn't notice Y" after mindfulness training?), and survey feedback on safety climate. For example, after introducing mindfulness and stress management training, a company might find that employees report improved concentration and teamwork. In AAM, one can track whether crews that undergo such training exhibit any measurable differences in performance during high-workload simulations or real operations (e.g., fewer missed checklist items, more stable task management). Even if the metrics are subtle, the combination of anecdotal feedback and any reduction in incidents would support the value of these human-centric safety investments.

CROSS-SECTOR COMPARISONS AND LESSONS FOR AAM

Throughout this chapter, we have interlaced comparisons with traditional aviation, manufacturing, and defense sectors. It is helpful to summarize a few salient cross-sector lessons about learning and resilience that AAM can leverage.

From Traditional Aviation

Perhaps the most directly relevant lessons come from the evolution of safety culture in commercial and general aviation. Aviation taught us the importance of reporting systems (e.g., NASA ASRS) and non-punitive environments to enable learning. It showed that rigorous incident investigations and global sharing of findings (via airworthiness directives, etc.) can prevent repeat accidents. CRM training demonstrated that human factors—teamwork, communication, and leadership—are just as critical as technical skills for safety. AAM should adopt these practices wholesale, while tailoring them to its specific needs. The SMS framework now mandated in aviation is essentially a codified learning system—AAM's novelty is not an excuse to skip SMS, but rather an impetus to implement SMS early and make it agile and data-driven. Another lesson is the shift toward performance-based regulation: aviation regulators are increasingly focusing on outcomes and continuous improvement, rather than just static compliance. For AAM, regulators are already considering performance-based criteria for urban air mobility operations, which means operators must continually learn and demonstrate improvement, rather than just meeting a one-time certification standard.

From Manufacturing (Industrial Safety and Quality)

The manufacturing industry, particularly sectors such as automotive and electronics, has developed matured continuous improvement systems, including Lean and Six Sigma (Çeliktaş et al. 2024; Hąbek et al., 2023; Sakib et al., 2025; Wong et al., 2009). These systems involve frontline workers in identifying problems (via Kaizen events or suggestion programs) and use data (statistical process control) to drive learning and quality improvements (Gejdoša, 2015; Męcfel, 2024; Salvadorinho et al., 2024). A parallel in AAM would be to empower all levels of staff—including pilots, maintenance technicians, software developers, and dispatchers—to suggest safety improvements and to view deviations as opportunities to refine processes (Salvadorinho et al., 2024). Manufacturing also shows how to integrate automation safely into workflows. In modern factories with collaborative robots, companies implement strict safety protocols (such as sensors and physical separation when needed) and also train workers to interact safely with robots (Geiger et al., 2025; Ghanbarzadeh & Najafi, 2025). They learned that both engineered safeguards *and* human awareness are required. Similarly, vertiport crews and maintenance staff in AAM will need to work safely around autonomous vehicles (for instance, loading passengers or cargo into a self-driving eVTOL); a combination of engineered interlocks and worker training will be necessary to ensure safety. Manufacturing espouses a *stop-the-line* authority, where any worker can halt production if they identify a safety issue or quality problem (Bell & Martinez, 2019). Translating that to AAM, any crew member or remote pilot should feel empowered to abort a mission or ground a vehicle if something seems unsafe. This cultural element ensures rapid response and learning, because investigating why someone hit the stop button often reveals issues that can be addressed systemically.

From Defense (Military Aviation and Operations)

Defense organizations operate in high-risk, dynamic environments and have institutionalized learning practices such as after-action reviews (AARs), conducted during and at the conclusion of training events and operations (Cates et al., 2022; Department of the Army, 2021; Mastaglio et al., 2011; Petrache, 2025). These AARs are often brutally honest, focusing on what went wrong and right, and involve all ranks—a model of a learning debrief that AAM teams could adopt after major test flights or operations. The military's concept of *Mission Command* grants subordinate leaders the freedom to adapt to changing circumstances, which parallels the need for remote pilots or local vertiport managers to have leeway to innovate solutions when unexpected events occur, with the expectation that they'll share lessons learned afterward (Department of the Army, 2012; Department of the Army, 2021; Department of the Army, 2022). HMT research in the military (like the trust calibration exercises mentioned earlier) is directly informative for AAM's remote operations and autonomous systems—in fact, some AAM vehicle concepts stem from military UAV designs, so leveraging that body of human factors knowledge (e.g., optimal operator workload, interface design for drone control centers) is wise (Chancey et al., 2020; Chen et al., 2025; Mayer, 2023). The defense sector also emphasizes redundancy and resilience: systems are designed to fail gracefully, and personnel are cross-trained to assume key roles. AAM should ensure that if, for instance, an AI-based dispatch system fails, humans can revert to a manual backup; drills should include operating in degraded modes (much as pilots practice flying with failed instruments or controllers simulate radio outages). These preparations instill confidence that the organization can handle surprises and learn from them, rather than being crippled by them.

From High-Reliability Organizations (HROs) in General

Industries such as nuclear power, petrochemicals, and healthcare have recently adopted Safety-II and resilience engineering concepts due to their need for ultra-high safety standards (Birkeli et al., 2025; Hamer et al., 2020; Hollnagel et al., 2015; Mohammed Ashraf et al., 2021). They emphasize *collective mindfulness*, as mentioned, and have experimented with Safety-II audits, where instead of looking only for things that went wrong, auditors observe work to see how and why things go right, capturing those successful adaptations as lessons. AAM regulators or internal safety teams might similarly conduct *positive audits*—for example, observing an AAM control center on a day when everything went smoothly and identifying which behaviors or decisions by staff ensured safety, then reinforcing those in training. This is analogous to FRAM's approach of analyzing normal performance variability. In healthcare, which faces both technical and human complexity, has also been highlighted the importance of EQ and effective communication in ensuring safety (AAOS, 2017; AAOS, 2015/2024; Codier & Codier, 2015). In surgical contexts, higher EQ is consistently framed as beneficial for nontechnical performance and the surgeon–patient relationship, and EI can be improved via targeted interventions; however, the literature linking EI directly to hard patient outcomes is still heterogeneous and developing. (Abi-Jaoudé et al., 2022; Rickard et al., 2023).

To visualize the concept of systemic learning, consider a Functional Resonance Analysis Method (FRAM) model of an AAM operation. FRAM helps map out the multiple functions and their interactions in a complex system, highlighting points of variability and potential areas for improvement. An example from a recent Formula 1 safety study shows a complex FRAM network of functions involved in risk assessment (Slater et al., 2021). Each hexagon in the diagram represents a function (with inputs, outputs, resources, controls, etc.), and the myriad connections show how functions depend on or influence each other. Even for a relatively structured process like the Formula 1 pit stop risk assessment (see Figure 7.3), the FRAM model looks like a web of interdependencies rather than a simple linear flow. By examining a FRAM diagram, safety teams can discuss "*What if X goes wrong? How will it impact Y and Z?*" and thereby anticipate potential resonant effects without having to experience every failure mode in reality.

AAM operations will likely be similarly complex with several nonlinear interactions. Tools like FRAM enable us to capture that complexity and engage in systemic learning. Rather than blaming a single component or person when something fails, we see the web of factors, analyze their interdependencies, and then be able to adjust the system to prevent future issues. For AAM, using such methods during design and testing phases (before full commercial rollout) could be extremely valuable, with a full cycle of resilience engineering to be applied, from anticipate, to monitor, to respond, and ultimately to learn, as already discussed in Chapter 4 (Figure 4.3) with the Work As Imagined (WAI) FRAM Model used in UAM airspace management scenario (see Figure 7.4). In short, cross-industry experience teaches AAM that proactively studying the system, through simulation, reporting, and analytical methods like FRAM, will pay off in safer operations.

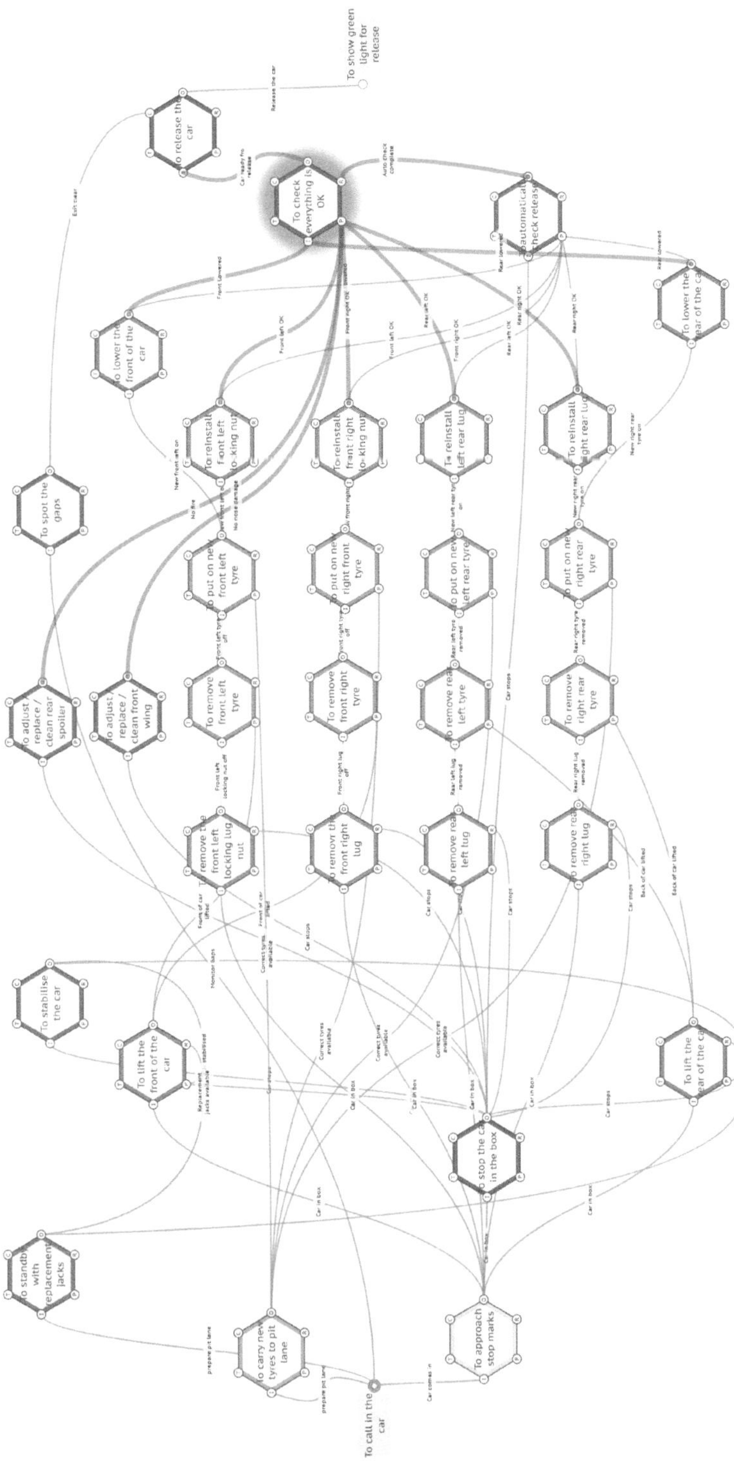

FIGURE 7.3 FRAM model for Formula 1 pit stop, work as done (WAD). (Credit: Slater et al., 2021; created with FRAM Model Visualizer.)

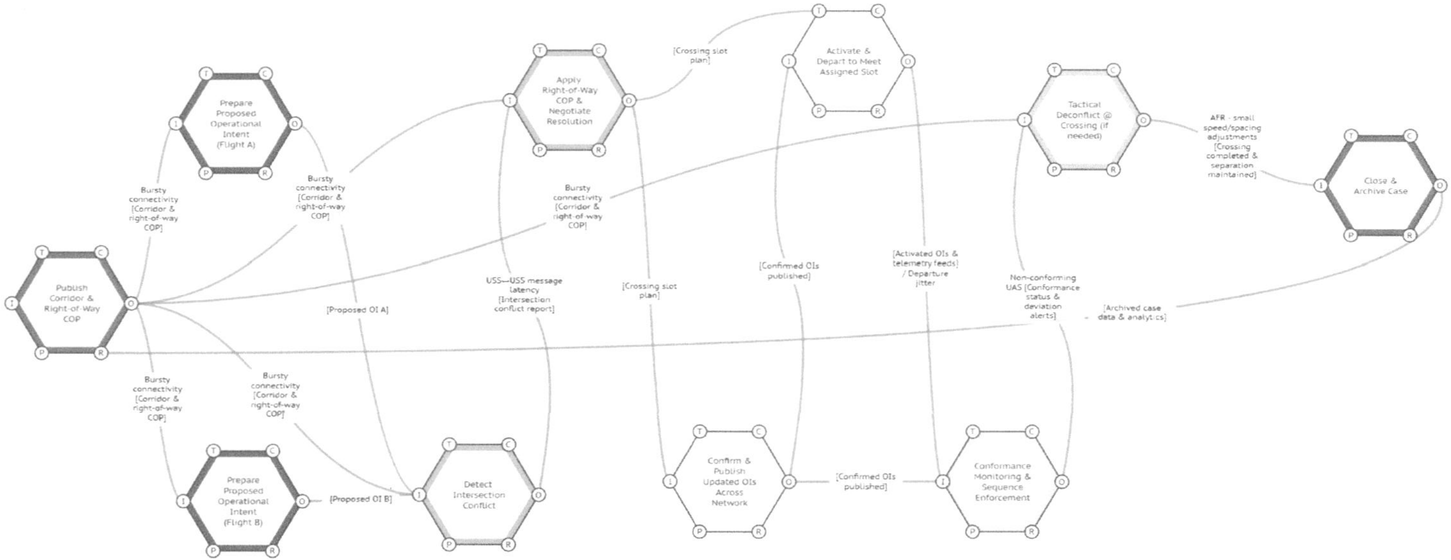

FIGURE 7.4 FRAM model for UAM airspace, work as imagined (WAI). (Credit: Author; created with FRAM Model Visualizer.)

REFERENCES

Abi-Jaoudé, J. G., Kennedy-Metz, L. R., Dias, R. D., Yule, S. J., & Zenati, M. A. (2022). Measuring and improving emotional intelligence in surgery: A systematic review. *Annals of Surgery, 275*(2), e353–e360. https://doi.org/10.1097/SLA.0000000000005022

Alaydi, B., & Ng, S. I. (2024). Mitigating the negative effect of air traffic controller mental workload on job performance: The role of mindfulness and social work support. *Safety, 10*(1), 20. https://doi.org/10.3390/safety10010020

American Academy of Orthopaedic Surgeons (AAOS). (2015; revised 2024). *Use of Structured Communication Tools to Improve Surgical Patient Safety* (Information Statement 1046).

American Academy of Orthopaedic Surgeons (AAOS). (2017, April 12). *Surgical team members need emotional intelligence, adaptability to prevent adverse events* (news release).

Artemenko, A. G. (2022). Human-autonomy teaming interface design for multiple-UAV control. *Russian Engineering Research, 42*(2), 418–424. https://doi.org/10.3103/S1068799822020222

Bates, S. (2023). Emotional intelligence and safety culture in business aviation (Doctoral dissertation, Walden University). https://scholarworks.waldenu.edu/cgi/viewcontent.cgi?article=15249&context=dissertations

Beattie, T., Pakpour, A. H., & Sharp, C. (2023). The association between emotional intelligence and decision making for pilots in the aviation industry. *Journal of Air Transport Management, 110*, 102149. https://doi.org/10.1016/j.jairtraman.2023.102506

Bell, S. K., & Martinez, W. (2019). Every patient should be enabled to stop the line. *BMJ Quality & Safety, 28*(3), 172–176. https://doi.org/10.1136/bmjqs-2018-008714

Birkeli, G., Lindahl, A. K., Hammersbøen, Å. M., Tveter Deilkås, E. C., & Ballangrud, R. (2025). Strategies and tools to learn from work that goes well within healthcare patient safety practices: A mixed methods systematic review. *BMC Health Services Research, 25*, 538. https://doi.org/10.1186/s12913-025-12680-2

Boyar, S. L., Savage, G. T., & Williams, E. S. (2023). An adaptive leadership approach: The impact of reasoning and emotional intelligence (EI) abilities on leader adaptability. *Employee Responsibilities and Rights Journal, 35*, 565–580. https://doi.org/10.1007/s10672-022-09428-z

Buck, B. K., Politowicz, M. S., Nelson, S. L., & Chancey, E. T. (2025). Preliminary development of multi-vehicle (m:N) operations with NASA Langley's Remote Vehicle Operations Center. In *Proceedings of AIAA SciTech 2025 Forum*. https://ntrl.ntis.gov/NTRL/dashboard/searchResults/titleDetail/N240014944.xhtml

Cates, K., Banghart, M., & Plant, A. (2022). Improving after action review (AAR): Applications of natural language processing and machine learning. *Journal of Military Learning, 6*(1), 3–13. Army University Press.

Çeliktaş, A., Alımlı, D., Haymour, M., & Tunaboylu, B. (2024). Application of DMAIC methodology in semiconductor manufacturing for yield improvement. In N. M. Durakbasa & M. G. Gençyılmaz (Eds.), *Industrial engineering in the Industry 4.0 era (ISPR 2023)* (pp. 262–274). Springer. https://doi.org/10.1007/978-3-031-53991-6_20

Chancey, E. T., Politowicz, M. S., & Le Vie, L. R. (2020). Enabling Advanced Air Mobility operations through appropriate trust in human-autonomy teaming: Foundational research approaches and applications (NASA TM–20205003378). NASA Langley Research Center. https://ntrs.nasa.gov/api/citations/20205002989/downloads/Chancey%20Politowicz%20Le%20Vie_SciTech2021(11.20.2020).pdf

Chen, D., Yoon, H. J., Wan, Z., Alluru, N., Lee, S. W., He, R., Moore, T. J., Nelson, F. F., Yoon, S., Lim, H., Kim, D. D., & Cho, J. H. (2025). *Advancing human-machine teaming: Concepts, challenges, and applications* (Version 2) [Preprint]. https://doi.org/10.48550/arXiv.2503.16518

Chevalier, S., Perot, A., Klenkenberg, S., Dumoulin, F., Declaye, J., & Stipulante, S. (2025). Emotional intelligence, transformational leadership, and team satisfaction during the COVID-19 period in Belgium. *Frontiers in Organizational Psychology*, *3*, 1578835. https://doi.org/10.3389/forgp.2025.1578835

Clarke, C. L. (2023). Power-up with emotional intelligence: Exploring flight attendant leadership and crew resource management in a dynamic post-pandemic travel environment (Doctoral dissertation, University of Southern Mississippi). https://aquila.usm.edu/dissertations/2195/

Codier, E., & Codier, D. (2015). A model for the role of emotional intelligence in patient safety. *Asia-Pacific Journal of Oncology Nursing*, *2*(2), 112–117. https://doi.org/10.4103/2347-5625.157594

Coronado-Maldonado, I., & Benítez-Márquez, M. D. (2023). Emotional intelligence, leadership, and work teams: A hybrid literature review. *Heliyon*, *9*(10), e20356. https://doi.org/10.1016/j.heliyon.2023.e20356

Dane, E. (2011). Paying attention to mindfulness and its effects on task performance in the workplace. *Journal of Management*, *37*(4), 997–1018. https://doi.org/10.1177/0149206310367948

de Visser, E., & Parasuraman, R. (2011). Adaptive aiding of human-robot teaming: Effects of imperfect automation on performance, trust, and workload. *Journal of Cognitive Engineering and Decision Making*, *5*(2), 209–231. https://doi.org/10.1177/1555343411410160

de Visser, E. J., Peeters, M. M. M., Jung, M. F., Kohn, S., Shaw, T. H., Pak, R., & Neerincx, M. A. (2020). Towards a theory of longitudinal trust calibration in human–robot teams. *International Journal of Social Robotics*, *12*, 459–478. https://doi.org/10.1007/s12369-019-00596-x

Department of the Army. (2012). *ADP 6-0: Mission Command* (17 May 2012). Headquarters, Department of the Army.

Department of the Army. (2021). *FM 7-0: Training* (incl. Appendix K: After Action Reviews). Headquarters, Department of the Army.

Department of the Army. (2022). *AR 11–33: Army Lessons Learned Program* (28 July 2022). Headquarters, Department of the Army.

Dugger, Z. T., & McCrory, B. (2021). Emotional intelligence and safety citizenship among Army aviators. *International Journal of Aviation, Aeronautics, and Aerospace*, *8*(1). https://doi.org/10.15394/ijaaa.2021.1552

Duplechan, L. (2024). How high reliability can facilitate clinical, organizational, and public health responses to global ecological health risks. *AMA Journal of Ethics*, *26*(2), E171–E178. https://doi.org/10.1001/amajethics.2024.171

Endsley, M. R., & Kiris, E. O. (1995). The out-of-the-loop performance problem and level of control in automation. *Human Factors*, *37*(2), 381–394. https://doi.org/10.1518/001872095779064555

Energy Institute. (2014). *Guidance on crew resource management (CRM) and non-technical skills training programmes* (1st ed., July 2014). Energy Institute, London.

FAA Safety Team & NASA. (2016). The case for confidential incident reporting systems. NASA ASRS. https://asrs.arc.nasa.gov/docs/rs/60_Case_for_Confidential_Incident_Reporting.pdf

Federal Aviation Administration (FAA). (2016). *Air Carrier Training Aviation Rulemaking Committee (ACT ARC) Recommendation 16-9: Manual Flight Operations.*

Federal Aviation Administration (FAA). (2017). *SAFO 17007: Manual Flight Operations Proficiency* (May 4, 2017).

Federal Aviation Administration (FAA). (2020). *Human Factors and Automation Interaction* (REDAC Full Committee briefing; July 2020).

Federal Aviation Administration. (2025a). Aviation Safety Action Program (ASAP). https://www.faa.gov/about/initiatives/asap

Federal Aviation Administration. (2025b). Temporary Flight Restrictions (TFRs). https://www.faa.gov/uas/getting_started/temporary_flight_restrictions

Geiger, L., Guadarrama-Olvera, J. R., & Cheng, G. (2025). Normative safety regulations for collaborative robots. In *Proceedings of the 2025 IEEE/SICE International Symposium on System Integration* (pp. 707–714). https://portal.fis.tum.de/en/publications/normative-safety-regulations-for-collaborative-robots/

Gejdoša, P. (2015). Continuous quality improvement by statistical process control. *Procedia Economics and Finance, 23*, 1186–1191. https://doi.org/10.1016/S2212-5671(15)01669-X

Geller, E. S. (2023). Emotional intelligence: A crucial human dynamic for OSH. *Professional Safety, 68*(12), 25–30. https://onepetro.org/PS/article-abstract/68/12/24/538660/Emotional-Intelligence-A-Crucial-Human-Dynamic-for?redirectedFrom=fulltext

Gemmano, C. G., Giancaspro, M. L., Galiotto, S., & Manuti, A. (2023). Resilience training for critical situation management: An umbrella and systematic review. *Safety Science, 168*, 106321. https://doi.org/10.1016/j.ssci.2023.106311

Gemmano, C. G., Giancaspro, M. L., Galiotto, S., & Manuti, A. (2024). A flight path to well-being: The mediating role of continuous learning between burnout and work performance in aviation professionals. *Social Sciences, 13*(10), 513. https://doi.org/10.3390/socsci13100513

Ghanbarzadeh, A., & Najafi, E. (2025). Safe physical human–robot interaction through variable impedance control based on ISO/TS 15066. *International Journal on Interactive Design and Manufacturing, 19*, 4741–4758. https://doi.org/10.1007/s12008-024-02074-9

Glomb, T. M., Duffy, M. K., Bono, J. E., & Yang, T. (2011). Mindfulness at work. *Research in Personnel and Human Resources Management, 30*, 115–157. https://doi.org/10.1108/S0742-7301(2011)0000030005

Gouraud, J., Delorme, A., & Berberian, B. (2017). Autopilot, mind wandering, and the out of the loop performance problem. *Frontiers in Neuroscience, 11*. https://doi.org/10.3389/fnins.2017.00541

Gregory, I. M., & Trujillo, A. C. (2016, July 5). *Self-aware vehicles for effective human–machine teaming* [Workshop presentation]. Collaborative Sensing, Learning, and Control in Human-Machine Systems Workshop at the 2016 American Control Conference, Boston, MA.

Hąbek, P., Lavios, J. J., & Grzywa, A. (2023). Lean manufacturing practices assessment: Case study of an automotive company. *Production Engineering Archives, 29*(3), 311–318. https://doi.org/10.30657/pea.2023.29.36

Hallé, S., Perrot, A., & Stipulante, S. (2025). Emotional intelligence training in health students: A pilot study. *Frontiers in Sports and Active Living, 7*, 1511851. https://doi.org/10.3389/fspor.2025.1511851

Hamer, R., Jun, G. T., & Waterson, P. (2021). Adaptations to everyday work amongst nuclear operators: A Safety-II approach. In R. Charles & D. Golightly (Eds.), *Contemporary ergonomics and human factors 2020* (pp. 1–9). Chartered Institute of Ergonomics & Human Factors. https://publications.ergonomics.org.uk/uploads/Adaptations-to-everyday-work-amongst-nuclear-operators-A-safety-II-approach.pdf

Hollnagel, E. (2009). The four cornerstones of resilience engineering. In C. P. Nemeth, E. Hollnagel, & S. W. A. Dekker (Eds.), *Resilience engineering perspectives, Volume 2: Preparation and restoration* (pp. 117–133). Ashgate.

Hollnagel, E. (2012). FRAM: The functional resonance analysis method – Modelling complex socio-technical systems. Ashgate. https://www.taylorfrancis.com/books/mono/10.1201/9781315255071

Hollnagel, E. (2014). *Safety-I and Safety-II: The past and future of safety management.* Ashgate (Routledge). https://www.taylorfrancis.com/books/mono/10.1201/9781315607511

Hollnagel, E. (2015). *Introduction to the Resilience Analysis Grid (RAG)* (Technical note).

Hollnagel, E., Wears, R. L., & Braithwaite, J. (2015). *From Safety-I to Safety-II: A White Paper.* The Resilient Health Care Net.

Johnson, C. J., Demir, M., McNeese, N. J., Gorman, J. C., Wolff, A. T., & Cooke, N. J. (2021). The impact of training on human–autonomy team communications and trust calibration. *Human Factors: The Journal of the Human Factors and Ergonomics Society, 65*(7), 1554–1570. https://doi.org/10.1177/00187208211047323

Joyner, P., & Lardner, R. (2008). Mindfulness: Realising the benefits. *Loss Prevention Bulletin, 201*, 22–27.

Lebares, C. C., Guvva, E. V., Olaru, M., Sugrue, L. P., Staffaroni, A. M., Delucchi, K. L., Kramer, J. H., Ascher, N. L., & Harris, H. W. (2019). Efficacy of mindfulness-based cognitive training in surgery: Additional analysis of the mindful surgeon pilot randomized clinical trial. *JAMA Network Open, 2*(5), e194108. https://doi.org/10.1001/jamanetworkopen.2019.4108

Lebiere, C., Blaha, L. M., Fallon, C. K., & Jefferson, B. (2021). Adaptive cognitive mechanisms to maintain calibrated trust and reliance in automation. *Frontiers in Robotics and AI, 8*, 652776. https://doi.org/10.3389/frobt.2021.652776

Lee, C.-C., Yeh, W.-C., Yu, Z., & Lin, X.-C. (2023). The relationships between leader emotional intelligence, transformational leadership, and transactional leadership and job performance: A mediator model of trust. *Heliyon, 9*(8), e18007. https://doi.org/10.1016/j.heliyon.2023.e18007

Lieberman, K. M. (2022). *Supporting Trust Calibration and Attention Management in Human-Machine Teams Through Training and Real-Time Feedback on Estimated Performance* (Doctoral dissertation, University of Michigan).

Luebbers, M. B., Tabrez, A., & Hayes, B. (2022). Descriptive and prescriptive visual guidance to improve shared situational awareness in human–robot teaming. In *Proceedings of AAMAS 2022* (pp. 1256–1264). https://m-luebbers.github.io/pdfs/aamas22.pdf

Masi, G., Amprimo, G., Ferraris, C., & Priano, L. (2023). Stress and workload assessment in aviation—A narrative review. *Sensors, 23*(7), 3556. https://doi.org/10.3390/s23073556

Mastaglio, T., Wilkinson, J., Jones, P. N., Bliss, J. P., & Barnett, J. S. (2011). *Current practice and theoretical foundations of the after action review* (Technical Report 1290). U.S. Army Research Institute for the Behavioral and Social Sciences.

Mayer, M. J. (2023). Trusting machine intelligence: Artificial intelligence and human–autonomy teaming in military operations. *Defense & Security Analysis, 39*(4), 521–538. https://doi.org/10.1080/14751798.2023.2264070

Męcfel, E. (2024). Case study: Development of employee suggestion system at OKNOPLAST manufacturing company as element of process improvement. *Decision Making in Manufacturing and Services, 18*(S.I), 49–58. https://doi.org/10.7494/dmms.2024.si.6690

Mezentseva, A., Gracia, F. J., Silla, I., & Martínez-Córcoles, M. (2023). The role of empowering leadership, safety culture and safety climate in the prediction of mindful organizing in an air traffic management company. *Safety Science, 168*, 106321. https://doi.org/10.1016/j.ssci.2023.106321

Mohammed Ashraf, A., Son, C., Peres, S. C., & Sasangohar, F. (2021). Navigating operating procedures in everyday work in a petrochemical facility: A comparative analysis of work-as-imagined and work-as-done. *Proceedings of the Human Factors and Ergonomics Society Annual Meeting, 65*(1), 623–627.

Moserle, G., Foti, G., Finstad, G. L., Bazzoli, A., Curcuruto, M., & Morgan, J. (2024). Prosocial behavior and workplace safety: The roles of emotional intelligence and perceived organizational support in the UK and US. *Sustainability, 16*(21), 9190. https://doi.org/10.3390/su16219190

NASA Aviation Safety Reporting System (NASA ASRS). (2024). System-Wide Safety: In-Time Aviation Safety Management System overview. https://ntrs.nasa.gov/citations/20240011545

NASA Aviation Safety Reporting System (NASA ASRS). (2025). UAS safety reporting (report set). https://asrs.arc.nasa.gov/docs/rpsts/uas.pdf

National Transportation Safety Board. (2022). NTSB aviation accident report no. 100950 [PDF]. https://data.ntsb.gov/carol-repgen/api/Aviation/ReportMain/GenerateNewestReport/100950/pdf

Neogi, N. A., Graydon, M. S., Holbrook, J. B., Maddalon, J. M., & McCormick, G. F. (2025). *Enabling novel collaborative control paradigms between humans and machines in electric vertical takeoff and landing (eVTOL) aircraft.* Presented at the Vertical Flight Society 81st Annual Forum & Technology Display, Virginia Beach, VA (May 20–22, 2025).

Nurkin, T., & Siegel, J. (2023). *Battlefield applications for human-machine teaming: Demonstrating value, experimenting with new capabilities, and accelerating adoption.* Atlantic Council (Scowcroft Center for Strategy and Security), August 2023. https://www.atlanticcouncil.org/wp-content/uploads/2023/08/Battlefield-Applications-for-HMT.pdf

Petrache, T. (2025). From cadet to commander: Leveraging after action review for leadership development in military academy. *International Conference Knowledge-Based Organization, 31*(1), 160–165. https://doi.org/10.2478/kbo-2025-0019

Rickard, M. J., Kozlowski, D., & Schnitzler, M. (2023). Outcomes of emotional intelligence training for surgeons in a real-world setting: A mixed methods study. *Journal of Surgical Education, 80*(10), 1445–1453. https://doi.org/10.1016/j.jsurg.2023.07.021

Robson, S., & Manacapilli, T. (2014). *Enhancing performance under stress: Stress inoculation training for battlefield airmen.* RAND Corporation. https://www.rand.org/pubs/research_reports/RR750.html

Sagirli, F. Y., Zhao, X., & Wang, Z. (2024). Human-computer interaction and human-AI collaboration in advanced air mobility: A comprehensive review. https://arxiv.org/abs/2412.07241

Sakib, M. N., Kawsar, M., & Bithee, M. M. (2025). Continuous improvement through Lean Six Sigma: A systematic literature review and bibliometric analysis. *International Journal of Lean Six Sigma, 16*(5), 1252–1275. https://doi.org/10.1108/IJLSS-08-2024-0173

Salvadorinho, J., Pintor, P., Bastos, T., & Teixeira, L. (2024). DisneyHint: Lean and coaching-based employee suggestion system for the human challenges of Industry 4.0. *Journal of Industrial Engineering and Management, 17*(2), 516–541. https://doi.org/10.3926/jiem.6314

Samu, J., Yang, C., & Winter, S. R. (2025). Assessing the role of emotional intelligence in risk behavior across safety critical environments: A systematic review. *Transportation Research Interdisciplinary Perspectives, 33*, 101627. https://doi.org/10.1016/j.trip.2025.101627

Shonin, E., & Van Gordon, W. (2014). Searching for the present moment. *Mindfulness, 5*, 105–107. https://doi.org/10.1007/s12671-013-0248-0

Slater, D., Hill, R., Kumar, M., & Ale, B. (2021). Optimising the performance of complex sociotechnical systems in high-stress, high-speed environments: The Formula 1 pit stop test case. *Applied Sciences, 11*(24), 11873. https://doi.org/10.3390/app112411873

Stroeve, S., Smeltink, J., & Kirwan, B. (2022). Assessing and advancing safety management in aviation. *Safety, 8*(2), 20. https://doi.org/10.3390/safety8020020

UK Civil Aviation Authority. (2016). *CAP 737: Introduction to Crew Resource Management (CRM) and Threat and Error Management (TEM).* Civil Aviation Authority.

Wang, Z., Jiang, Z., & Blackman, A. (2023). Why and when do emotionally intelligent employees perform safely? The roles of thriving at work and career adaptability. *Applied Psychology.* Advance online publication. https://doi.org/10.1111/apps.12497 [atlanticcouncil]

Weick, K. E., & Sutcliffe, K. M. (2007). *Managing the unexpected: Resilient performance in an age of uncertainty* (2nd ed.). Wiley.

Weick, K. E., Sutcliffe, K. M., & Obstfeld, D. (1999). *Organizing for high reliability: Processes of collective mindfulness.* In R. S. Sutton & B. M. Staw (Eds.), *Research in Organizational Behavior* (pp. 81–123). JAI Press.

Wong, Y. C., Wong, K. Y., & Ali, A. (2009). *A study on lean manufacturing implementation in the Malaysian electrical and electronics industry. European Journal of Scientific Research*, *38*(4), 521–535.

You, F., Yan, X., Zhang, J., & Cui, W. (2022). Design factors of shared situation awareness interface in human–machine co-driving. *Information*, *13*(9), 437. https://doi.org/10.3390/info13090437

Part III

Sustainable Operations

In the third part of this publication, the spotlight shifts to the tangible achievements gained through the adoption of AAM technologies, which contribute to the goal of achieving a sustainable future for humanity across environmental, societal, and financial dimensions.

First, AAM innovations such as electric propulsion and advanced design can shrink aviation's carbon footprint and reduce noise pollution. For example, Zipline's battery-powered delivery drones emit 99% less carbon than diesel vans (Chatterjee, 2025), illustrating the potential for cleaner skies through the use of electrification. Similarly, many eVTOL developers envision quiet, carbon-neutral aircraft as a sustainable alternative to today's fuel-burning, noisy helicopters. These technologies thus promise substantial environmental gains if powered by renewable energy, helping aviation reduce its climate impact.

The human impact and community implications of AAM can be immense. These vehicles could revolutionize how people and goods move, from delivering medical supplies to remote communities to ferrying passengers above congested roads, in ways that improve quality of life. Real-world pilot programs already demonstrate such benefits: in Canada, drones are now transporting medicine to First Nations clinics in hard-to-reach areas (Flemons et al., 2022), and in India, the postal service used drones to cut a mountainous delivery route from 90 minutes by road to just 15 minutes by air (Aggarwal et al., 2023). By creating "highways in the sky," eVTOL air taxis might also alleviate urban traffic congestion (Sarkar et al., 2022), offering faster commutes and on-demand services. However, public acceptance will depend on factors such as safety, noise, and equitable access. Notably, noise has been a major hurdle for traditional helicopters; however, the quiet nature of battery-powered aircraft gives eVTOLs a

DOI: 10.1201/9781003546306-10

chance to succeed where louder aircraft have failed—a key consideration for gaining community and regulatory approval.

Finally, the economic sustainability and industry viability of this aerial revolution could be exponential. Although building new drone and eVTOL networks demands significant upfront investment, the long-term outlook is increasingly promising. Electric drones can reduce delivery costs by an estimated 40–70% compared to truck-based delivery (Chatterjee, 2025), mainly by saving on fuel, labor, and travel time. Major companies are already validating the model. Walmart, for instance, now deploys drones that can reach roughly 1.8 million U.S. households within 30 minutes, thereby avoiding the tailpipe emissions of equivalent van routes (Li et al., 2024). Amazon's Prime Air is aiming even higher, publicly targeting 500 million autonomous parcel flights per year by 2030 (Gordo Arias et al., 2024). Investor confidence mirrors this momentum: by 2024, the advanced air mobility sector encompassed more than 400 startups globally, while major industry trackers in early 2022 already counted approximately 250 eVTOL-focused companies—far more than any plausible long-run market could sustain (Lewis-Kraus, 2024; Zorpette & Ackerman, 2022). Several developers reached *unicorn* valuations (higher than $1B) well before routine commercial passenger operations (Swartz & Hirschberg, 2020; Technical University of Munich, 2020).

8 Environmental Implications

INTRODUCTION

Advanced air mobility (AAM)—encompassing electric air taxis, delivery drones, and other novel aerial vehicles—promises to transform transportation with unprecedented speed and convenience. These innovations conjure images of whisking above traffic jams in quiet, battery-powered "flying cars." Yet as exciting as this future is, a central question remains: can we achieve the benefits of AAM in a truly sustainable way? AAM operations must be evaluated not just for safety and efficiency, but also for their impact on our planet. In this chapter, we examine the environmental dimension of AAM on a global scale. We explore *what* AAM might achieve for environmental sustainability if deployed thoughtfully, and *how* to ensure this new mode of transport is an ecologically responsible part of future aviation. Key topics include greenhouse gas (GHG) emissions and climate impact, noise pollution and community noise exposure, energy sources (electric batteries, hydrogen fuel cells, and hybrids), lifecycle sustainability (from vehicle production to end-of-life disposal), land use and ecosystems, airspace congestion, and operational strategies to mitigate environmental footprint. Throughout, we present current technologies, real-world data, and case studies from leading programs, manufacturers, and cities piloting AAM solutions. The goal is to assess how AAM can be integrated into transportation in an environmentally sustainable manner as part of resilient aviation operations.

Achieving resilient safety in AAM is closely tied to achieving environmental sustainability. AAM operations cannot be considered truly safe or resilient if they cause long-term harm to the environment or are widely rejected by communities due to noise or other nuisances. The vision that emerges from our exploration is one of an AAM ecosystem that is clean, quiet, and energy-smart by design. This entails a framework that integrates vehicle innovation (electric and hydrogen propulsion), renewable energy infrastructure, advanced traffic management, and community-centric planning. If we keep environmental principles at the core of AAM development, the result could be a new mode of transportation that not only enhances mobility and emergency response (social benefits covered in Chapter 9) and creates new economic opportunities (financial aspects in Chapter 10), but also helps cities meet climate goals and improve quality of life. The sky can indeed be the limit for AAM—provided we approach the skies with a commitment to being good stewards of our planet. The early signs are encouraging: the pioneers of AAM are not ignoring sustainability, but rather embracing it as a key selling point and responsibility. As AAM moves from concept to reality in the coming decade, continuing this commitment will be essential. AAM presents a unique opportunity

DOI: 10.1201/9781003546306-11

to become a shining example of technology serving society sustainably, enabling the dream of flight in our daily lives while keeping our feet firmly on the ground in caring for Earth.

GREENHOUSE GAS (GHG) EMISSIONS AND CLIMATE IMPACT

One of the most frequently touted benefits of AAM is the potential to reduce GHG emissions by replacing or augmenting traditional fossil-fueled transportation with electrified flight. Electric vertical takeoff and landing (eVTOL) aircraft produce zero exhaust at the point of use, eliminating direct CO_2 and pollutant emissions during flight. This is in stark contrast to helicopters or small planes burning aviation fuel, as well as gasoline-powered cars on the ground. For example, an independent evidence review in the United Kingdom noted that AAM has the *"potential to reduce greenhouse gas emissions"* but also cautioned that the true impact is highly dependent on multiple variables, especially the energy mix used to recharge or fuel these vehicles (BryceTech, 2023). If the electricity powering eVTOL fleets comes from renewable sources (such as solar, wind, or hydrogen), the net carbon emissions can be significantly lower than those of gasoline-burning vehicles. On the other hand, if eVTOL charging relies on electricity generated from coal or gas, the emissions are simply shifted upstream to power plants. In short, not all electrons are green—the climate benefit of AAM hinges on decarbonizing the energy supply these advanced aircraft rely on. Therefore, policymakers and operators recognize this and are emphasizing the integration of clean energy as a prerequisite for AAM's climate credentials.

Lifecycle Emissions

A comprehensive view of AAM's climate impact must include lifecycle emissions—from manufacturing the vehicles and batteries, to building infrastructure, to end-of-life disposal. The production of high-performance AAM vehicles involves energy-intensive processes (e.g., curing carbon-fiber composite airframes and manufacturing lithium-ion battery cells), which incur an initial carbon footprint before the vehicle ever takes flight. Likewise, constructing vertiports and charging facilities incurs environmental costs (from concrete, steel, and land clearing) that should be amortized over the system's lifespan. For instance, a NASA analysis notes that the benefits of electric flight can be offset by the carbon and energy investment of building new aircraft and infrastructure if not managed sustainably (Schiller, 2023). During operations, all-electric eVTOL vehicles produce no tailpipe emissions, but there are indirect emissions if the electricity source is carbon-based. Hybrid-electric concepts would involve some combustion emissions in use (from a fuel-burning engine or generator) in addition to emissions associated with electricity generation for the hybrid system. Ultimately, the disposal and recycling of components, particularly large battery packs, will determine whether retired vehicles generate e-waste or can be integrated into a circular economy.

It is imperative that sustainable AAM necessitates planning for battery recycling and the use of recyclable or renewable materials in production. Many AAM stakeholders are exploring *green* manufacturing and end-of-life strategies, such as utilizing renewable energy in factories, designing components for disassembly, and establishing programs for battery second-life or recycling (Li et al., 2020; Rettenmeier et al., 2024; Tian & Chen). In short, the climate equation for AAM must account for *all* phases of the vehicle's lifecycle, not just the emissions (or lack thereof) while flying.

Comparisons with Ground Transport

A key question is how the emissions of AAM vehicles compare to those of the ground vehicles they might replace. Studies indicate that the answer depends on trip distance and passenger load. Kasliwal et al. (2019) performed an analysis of eVTOL energy use and GHG emissions versus cars. For a point-to-point trip of approximately 100 km carrying a single occupant, a representative eVTOL was found to emit 35% less GHG than a gasoline car, but about 28% more GHG than an electric car for the same trip. This less favorable comparison with a battery-electric ground vehicle is because short flights expend significant energy during vertical takeoff and climb. A single-passenger eVTOL doesn't spread that energy cost over multiple people. However, the equation improves markedly with higher occupancy and longer distances. With a fully loaded eVTOL (e.g., three passengers), the study suggested that on the same 100 km trip, per-passenger GHG emissions were 52% lower than those of a gasoline car and even about 6% lower than those of an electric car. In other words, high passenger capacity utilization and longer ranges can make eVTOL aircraft more *carbon-efficient per passenger-kilometer* than cars, especially when compared to typical single-occupant car travel. Recent case studies reinforce this point. A 2025 analysis for Paris found that on a 50 km urban route, eVTOLs would still produce slightly more CO_2 per passenger than a petrol car: roughly 13% higher, given the energy needed for vertical flight (Hagag & Hoeveler, 2025). However, on a 300 km regional route, a winged eVTOL could emit about 46% less CO_2 per passenger than a petrol car. Notably, in that study, the eVTOL still emitted more than an electric ground vehicle or train on the same route, by a factor of 68–96%, underlining that electric ground transport remains the gold standard for efficiency when available.

Furthermore, Figure 8.1 indicates hydrogen-eVTOL has the lowest lifecycle GHG (~31.5 g CO_2-eq/pax·km), battery-electric (BEV) cars average ~63 g, and battery eVTOLs (~70 g) are roughly on par with a turboprop commercial aircraft (~71 g).

The takeaway is that AAM's carbon efficiency is scenario-dependent: for very short trips, ground battery-electric vehicles may be superior, whereas for longer trips or when carrying multiple passengers, eVTOLs can outperform internal combustion vehicles in GHG per passenger-kilometer. One evidence review suggested a crossover point around 50 km: for trips longer than roughly 50 km, eVTOLs begin to outshine gasoline cars in efficiency, whereas for very short hops, ground EVs remain cleaner (BryceTech, 2023). Ensuring eVTOLs fly near full capacity (rather than mostly empty) will be critical to their sustainability advantage.

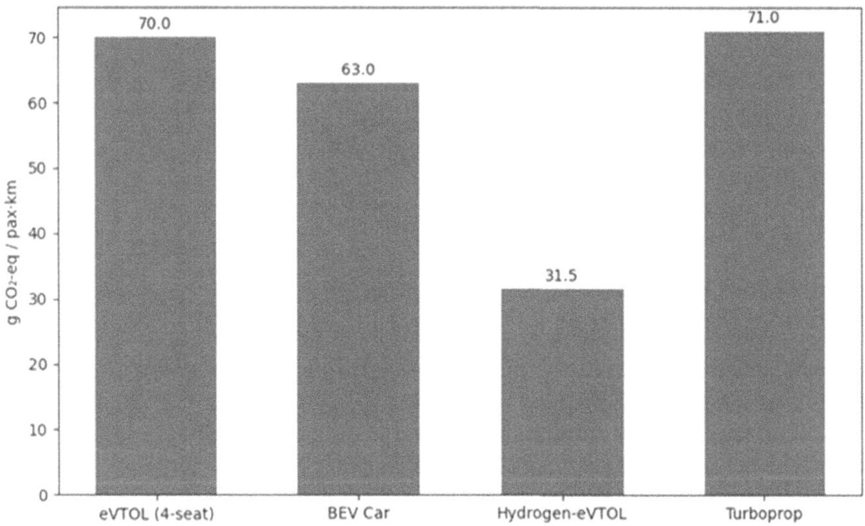

FIGURE 8.1 Lifecycle GHG per passenger-kilometer. (Credit: Author, Data source: Khavarian & Kockelman, 2023; Kasliwal et al., 2019; Song et al., 2025; Velaz-Acera et al., 2025.)

Replacing High-Emitting Modes

AAM's environmental value is strongest when it displaces high-emission modes, such as traditional helicopters or long solo car trips, rather than competing with already efficient options. Electric air taxis produce no operational CO_2 or NOx, whereas a helicopter of similar size emits significant CO_2 and other pollutants by burning Jet-A fuel. Even short-range commuter flights in eVTOLs could cut emissions if they replace driving in heavy congestion (where cars waste fuel idling) or eliminate the need for short, inefficient regional flights. For instance, a case study in South Korea estimated that 30,000 daily urban air mobility trips in the Seoul metro area could reduce CO_2 emissions by about 90,000 tons per year by supplanting some ground vehicle travel (Cho & Kim, 2022). Likewise, delivery drones and cargo eVTOLs have shown potential for emission savings: one analysis found that a city-wide drone delivery service could eliminate 94–95% of CO_2 emissions annually by reducing diesel truck mileage, especially for small, lightweight deliveries (Rodrigues et al., 2022). On a larger scale, a 2022 PwC study projected that by 2030, drones (across use cases) could deliver a 2.4 million-ton reduction in CO2e in the United Kingdom—equivalent to taking around 1.7 million diesel cars off the road for a year (PwC, 2022).All the above figures highlight how AAM, if deployed strategically, can make a meaningful contribution to transportation decarbonization, especially in niche markets where it replaces *inefficient* ground journeys or older, less environmentally friendly aircraft.

Electricity Demand and Grid Impact

As AAM scales up, charging potentially hundreds of electric aircraft will add load to power grids, raising an important environmental consideration beyond just vehicle efficiency. If unmanaged, a surge in electricity use for eVTOL charging could stress local grids and even trigger increased use of standby fossil fuel plants (peaking power generators), potentially eroding the environmental gains. To avoid simply shifting emissions upstream, strategies are being developed to power AAM with clean energy and to smooth out the impact on the grid. One approach is integrating on-site renewable energy generation and battery storage at vertiports. For example, some prototype vertiport designs incorporate solar panels and large battery banks, enabling eVTOLs to be charged partially or entirely off-grid using renewable energy. The *Air One* vertiport demonstrator in Coventry, UK, in 2022 featured a microgrid with solar photovoltaics and hydrogen fuel cell backup, showcasing how future eVTOL hubs might operate on green energy independent of the legacy grid (Sandhu, 2024). Another approach is smart charging, which involves scheduling eVTOL charging for off-peak times or when the renewable energy supply is high. Fleet operators could time most charging for midday (solar peak) or overnight (when wind energy or overall demand patterns make grid electricity cleaner and cheaper).

Drawing on NASA reports and acoustic-propagation models, indicated in Figure 8.2, the analysis concludes that eVTOLs are ~15 dB quieter than piston helicopters at ~300 m and should be routed using contour-based, noise-aware methods informed by spherical spreading, while energy–distance curves show a pronounced short-trip "hover penalty" (<≈35 km) where hover power dominates per-kilometer demand, using parameters from recent eVTOL energy and battery studies.

Emerging studies by the U.S. National Renewable Energy Laboratory (NREL) indicate that a busy vertiport might require on the order of one megawatt or more of power capacity—roughly equivalent to the peak demand of a large retail center—to fast-charge multiple eVTOLs simultaneously (Zart, 2024). This level of demand may necessitate local grid upgrades or creative solutions, such as energy storage, to enable the vertiport to draw a steady, manageable load from the grid and supplement it with batteries during peak charging periods. If done right, AAM's electricity

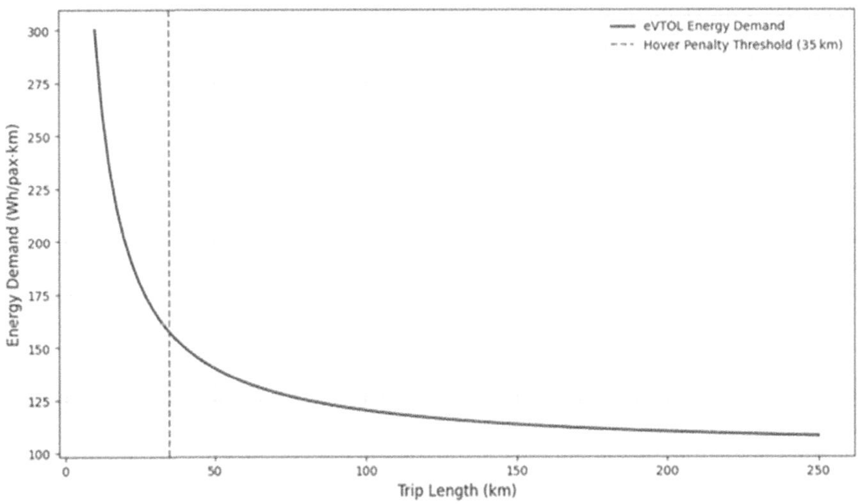

FIGURE 8.2 eVTOL energy demand vs. trip length.

demand could be met with minimal emissions. Policy measures such as incentivizing renewable electricity for AAM and setting carbon intensity standards for eVTOL operations may also be used to ensure AAM truly advances climate goals. Ultimately, close coordination between the aviation sector and energy providers, along with robust sustainability policies, will be needed to tilt the outcome toward the beneficial end of that spectrum.

NOISE POLLUTION AND COMMUNITY NOISE EXPOSURE

Apart from climate concerns, noise pollution is perhaps the most immediate environmental issue facing AAM deployment. Urban air mobility vehicles will operate at low altitudes near where people live and work. Hence, the noise they generate is crucial for public acceptance and for meeting environmental noise regulations. Conventional helicopters are notoriously loud, which has historically limited their acceptance in cities. In response, AAM developers are engineering eVTOLs to be significantly quieter than helicopters, leveraging the inherent advantages of electric propulsion: the absence of combustion noise and the ability to utilize multiple smaller rotors with optimized speeds and blade designs. Many eVTOL designs aim for a 15–20 dB noise reduction compared to helicopters of similar size (ICAO, 2022). A 20 dB reduction is significant—in simple terms, it would shrink the radius of a given noise contour to about one-tenth the distance, covering roughly 1% of the area of the helicopter's noise footprint (ICAO, 2022). Practically, this means far fewer people would be exposed to high noise levels. Even a 10 dB noise cut is substantial, equating to roughly a halving of perceived loudness to the human ear. So, the prospect of eVTOLs that are tens of decibels quieter than helicopters represents a transformative improvement in community noise impact.

Early flight tests are indeed validating that leading eVTOL designs can achieve extremely low noise signatures. In 2021–2022, NASA and Joby Aviation conducted acoustic flight tests of Joby's five-seat eVTOL prototype as part of NASA's AAM National Campaign. The results were groundbreaking: when flying in winged cruise mode at ~100 knots and 500 m (about 1,600 ft) above ground, the eVTOL produced only about 45 dB(A) of sound as measured on the ground (Blain, 2022). For context, 45 dB(A) is roughly equivalent to the background noise level of a quiet suburban neighborhood or the sound of a refrigerator humming in the next room. In other words, the Joby aircraft cruising overhead was barely perceptible against typical urban ambient noise. Even during the more energy-intensive vertical takeoff and landing phases, the measured noise 100 m from the vertiport remained below 65 dB(A)—about the level of a normal conversation or a vacuum cleaner at 100 m distance. These figures are dramatically lower than those for a traditional helicopter of similar size. For comparison, Figure 8.3, a small helicopter typically registers on the order of 85–90 dB(A) at 100 m during takeoff or flyover (Orikpete et al., 2020), which is why helicopter operations can be particularly disruptive over populated areas.

Moreover, the character of the noise from eVTOLs like Joby's is different from that of helicopters. Observers describe it as a benign, low-frequency hum or "whoosh" that tends to blend into city soundscapes, lacking the sharp blade-slap or

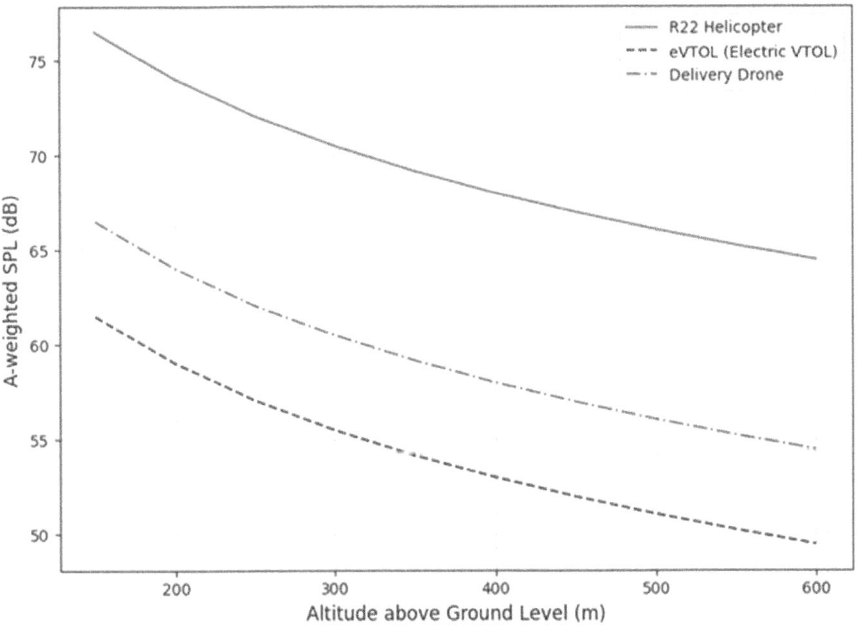

FIGURE 8.3 A-weighted noise footprint (dB) by altitude. (Credit: Author, Data source: Acoustic Analytics, n.d.; Discovery of Sound in the Sea, n.d.; Engineering ToolBox, n.d.; European Union Aviation Safety Agency, 2025; Gao et al., 2024; Rizzi et al., 2020; Rizzi, 2022; Rizzi et al., 2023; Schäffer et al., 2021; Schiller, 2023.)

thumping sounds of a helicopter (Blain, 2022). This more neutral tone means that, even if an eVTOL is audible, it may be perceived as less annoying than the high-pitched whine of a small drone or the "thwack-thwack" of rotors. The International Civil Aviation Organization (ICAO) has noted that such improvements (15–20 dB lower noise than helicopters) could reduce by orders of magnitude the number of people affected by aircraft noise in urban areas (ICAO, 2022). However, ICAO also highlights the need to develop appropriate noise metrics and certification standards tailored to these new types of aircraft. Traditional noise metrics may not fully capture human annoyance for eVTOL sound profiles; therefore, regulators are developing new methods to measure and limit eVTOL noise in a way that protects communities.

Of course, a single eVTOL flight that is barely noticeable is one thing, but what about hundreds of flights per day crisscrossing a city? The cumulative impact of many AAM vehicles is an area of active research and a valid public concern. A single drone or air taxi passing overhead might not bother most people, but a constant stream of them could create a new layer of ambient noise if not managed carefully. NASA, for instance, is developing tools for UAM *fleet noise* prediction and is studying the *dose-response* relationship—how a certain number of overflights per day at given noise levels might translate into community annoyance or disturbance (Schiller, 2023). Metrics such as Community Noise

Equivalent Level (CNEL) or Day-Night Average Sound Level (DNL), which aggregate noise exposure over time, are being considered to quantify the overall AAM noise impact in an area (City of Los Angeles, 2021). Early results suggest that even if individual AAM aircraft are quiet, operational constraints may be needed to prevent noise pollution hotspots in neighborhoods. Strategies under discussion include limiting flights during sensitive times (e.g., late at night or early morning), spacing flight paths to avoid concentrating traffic in one area, and designing approach/departure routes that minimize the impact on residential zones. Los Angeles, for example, is proactively involving communities in the AAM noise planning process. The city's Urban Movement Labs and transportation department created an immersive *Sound Lab* experience, playing recordings of eVTOL and helicopter sounds for residents to gauge reactions (City of Los Angeles, 2021). Using these simulations, Los Angeles found that specific sound frequencies and patterns were more acceptable and identified preferred corridors (mostly along existing noisy infrastructure, such as freeways and the LA River), where eVTOLs could fly with minimal new disturbance. This community-driven approach led to the proposed network of UAM routes, which avoid quiet residential areas and instead follow noisier rights-of-way, as well as recommendations on vertiport siting (e.g., co-locating vertiports with transit hubs or commercial zones rather than in the middle of neighborhoods). Such efforts demonstrate how public engagement and strategic planning can mitigate noise impacts before large-scale AAM operations commence.

It is also important to note that noise impacts are not limited to human communities—they also affect urban wildlife and ecosystems. Birds, in particular, can be sensitive to aircraft noise, which may disrupt feeding or nesting behaviors. Quieter eVTOLs will help, but the introduction of many low-altitude flights could still alter natural soundscapes if not carefully controlled. Parks and green spaces might experience increased noise if AAM corridors pass overhead, potentially affecting bird breeding or migratory patterns. As a precaution, AAM flights are routed along existing noisy corridors (such as highways or train lines) to minimize incremental noise in quiet areas. Environmental regulators will certainly factor in wildlife disturbance when reviewing new AAM routes or vertiport projects. In the United States, for example, any federal approval of AAM operations or infrastructure will undergo an environmental assessment under the National Environmental Policy Act (NEPA), which requires consideration of impacts on wildlife and natural quiet in parks (FAA, 2023). In Europe, a 2021 study by EASA on UAM societal acceptance found that noise was the top environmental concern among citizens, and it recommended that initial AAM routes avoid sensitive sites like hospitals and nature reserves and that aircraft use noise-abatement flight profiles to reduce impact (EASA, 2021).

ENERGY SOURCES: ELECTRIC, HYDROGEN, AND HYBRID PROPULSION

The environmental performance of AAM vehicles is inextricably linked to the energy sources and propulsion technologies they use. In broad terms, current and emerging AAM powertrains can be categorized into three main types: battery-electric, hydrogen fuel cell, and hybrid-electric. Each has its own set of environmental pros and cons.

Battery-Electric Propulsion

Most AAM prototypes and early models are fully battery-electric, meaning they utilize rechargeable lithium-ion battery packs to power electric motors that drive propellers or rotors. The appeal of battery-electric propulsion lies in its zero operational emissions and high energy efficiency. Electric motors are incredibly efficient, with an efficiency range of 80–90% (Kaya et al., 2021). Unlike combustion engines, their efficiency does not sharply degrade at smaller scales, which enables distributed propulsion (using multiple small motors) without losing efficiency. Indeed, one analysis notes that electric motors are about 2–3 times more efficient than comparable internal combustion engines, and using many small motors can reduce aerodynamic drag and improve overall aircraft efficiency (Sripad & Viswanathan, 2021). From an environmental standpoint, battery-electric AAM vehicles emit no direct CO_2 or NOx during flight. They also avoid the risk of fuel spills or combustion byproducts that come with hydrocarbon fuels. Importantly, electric power allows for much quieter operation.

However, the environmental challenges for battery-electric AAM are twofold. First, battery production and end-of-life pose significant environmental impacts. Mining and processing materials such as lithium, cobalt, nickel, and graphite can lead to habitat destruction and pollution. Disposing of or recycling batteries on a large scale remains a developing industry (Huang et al. 2025; Kang et al., 2013). Second, electricity generation for charging the batteries may involve carbon emissions if the grid is not green. Charging eVTOLs on a coal-heavy grid would transfer emissions to an off-site location. Thus, to maximize the sustainability of electric AAM, stakeholders are focused on both clean energy supply and responsible battery lifecycle management.

On the energy side, vertiports are being envisioned as green energy hubs. For example, the company Urban-Air Port (now renamed Urban-Air Choice) has designed off-grid vertiport systems that integrate solar panels, battery energy storage, and even hydrogen fuel cells on-site to supply clean electricity for eVTOL charging. In 2022, their prototype Air One vertiport in Coventry, UK, demonstrated this concept: the vertiport's microgrid provided power, showing how future eVTOL hubs might operate largely independent of fossil-fueled grids (Sandhu, 2024).

On the battery side, planning is underway for second-life use and recycling. eVTOL operators expect to replace battery packs periodically (perhaps after a few hundred or thousand cycles when capacity fades); those used batteries could be repurposed for stationary storage or recycled to recover materials (Leon & Miller, 2020; Spindlegger et al., 2025).

Hydrogen Fuel Cells

Hydrogen propulsion offers an alternative zero-emission solution by using hydrogen fuel cells to generate electricity on board the aircraft. In a fuel cell, hydrogen

(H_2) reacts with oxygen to produce electric power, with water vapor as the only direct emission (Aminudin et al., 2022). Hydrogen as a fuel has a very high specific energy (energy per unit weight) compared to current lithium batteries, which makes it attractive for improving the range and payload capacity of AAM vehicles. A prominent example is the *Alaka'i Technologies Skai* eVTOL prototype—a hydrogen fuel cell-powered air taxi designed to carry up to five people (Weitering, 2024). The Skai is expected to have an initial range of around 200 miles (320 km) on a tank of hydrogen, far beyond the typical 30–60 mile ranges being targeted by battery eVTOLs, precisely because hydrogen stores much more energy for the same weight. In mid-2019, Alaka'i conducted an unmanned flight demonstration of the Skai, highlighting its six-rotor design powered by onboard fuel cells (Weitering, 2024). The environmental advantage during operation is clear: no carbon emissions or air pollutants are emitted during flight, and the fuel cells and electric motors are as quiet as those of purely battery-powered systems.

Nevertheless, hydrogen's sustainability depends heavily on how the hydrogen is produced. Today, most industrial hydrogen is produced by reforming natural gas, a process that emits CO_2; this is known as *gray* hydrogen (Hermesmann & Müller, 2022). For hydrogen-powered AAM to be truly green, it would require green hydrogen (produced via electrolysis using renewable electricity) or, at the very least, blue hydrogen (from natural gas with carbon capture and storage to trap emissions). Generating hydrogen also consumes a significant amount of energy; some analysts argue that unless there is an abundant surplus of renewable energy, it may be more efficient to use that electricity to charge batteries directly than to convert it to hydrogen and then back to electricity via fuel cells (Pellow et al., 2015).

Another challenge is infrastructure: vertiports would require facilities for hydrogen storage and refueling, which raises safety considerations, as hydrogen is flammable and typically necessitates high-pressure tanks or cryogenic liquid storage (Yoo et al., 2021). Despite these challenges, there is significant investment and interest in hydrogen aviation due to its promise for longer-range, heavier-duty missions that batteries cannot yet economically support. Japan, for example, has expressed interest in hydrogen fuel cell options for future AAM vehicles, aiming to extend the range beyond what battery-only designs can achieve. While Japan's SkyDrive SD-05 eVTOL, being demonstrated at Expo 2025 in Osaka, is battery-electric, some Japanese consortia are exploring hydrogen as a power source for the next generation of inter-city vehicles (SkyDrive, 2025). Europe and the United States are also funding research into hydrogen propulsion for regional aviation, such as 19-seat hydrogen-electric commuter planes (Marksel & Brdnik, 2023). These developments could potentially trickle down to AAM. If hydrogen infrastructure becomes available—perhaps co-located at major vertiports or existing airports—hydrogen eVTOLs, like Skai, could serve routes of a few hundred kilometers with only water vapor as exhaust.

Battery-electric and hydrogen fuel cell propulsion will likely *complement* each other in the AAM ecosystem: battery eVTOLs may dominate short urban routes (where trips are brief and charging infrastructure is readily available). In contrast, hydrogen fuel cell eVTOLs could handle longer, suburban, or regional routes where extra range is needed and where hydrogen refueling can be centralized. Both are zero-emission at the point of use, so together they could enable AAM to cover a wide range of missions without tailpipe emissions, as long as the electricity or hydrogen fuel comes from clean sources.

Hybrid-Electric Systems

A hybrid AAM vehicle utilizes a combination of battery-powered electric motors and a conventional fuel-burning engine, typically a small turbine or piston engine that drives a generator, to extend range or provide redundancy. Early AAM concepts often leaned toward hybrid designs as a pragmatic step before full electrification. For example, the Bell Nexus air taxi concept, initially unveiled in 2019, was envisioned as a hybrid-electric vehicle (Johnson, 2020). It featured six tilting ducted fans powered by a turboshaft engine coupled to a generator, supplemented by batteries for extra power during takeoff and landing.

The environmental profile of hybrids is mixed. On the one hand, a hybrid eVTOL still burns fuel and emits CO_2 during operation, unlike a pure electric vehicle. On the other hand, the combustion engine in a hybrid can be run at a steady, efficient output (charging the battery or directly powering rotors), which is more fuel-efficient and cleaner than the highly variable, stop-and-go thrust profile of a helicopter. Because the electric side of the hybrid handles the energy-intensive bursts (such as vertical takeoff), the engine can be smaller and used primarily for cruise or to recharge batteries, resulting in reduced total fuel burn and emissions compared to a purely engine-driven aircraft of similar capability (Friedrich & Robertson, 2015; Zhang et al., 2022). Hybrids thus offer a bridge: they can fly longer or carry more weight than current battery-only designs, using today's fuel infrastructure, while still benefiting from some of the efficiency and noise advantages of electric drive. They also provide a backup power source (the engine) in case of battery depletion or failure, which some argue adds an extra layer of safety for specific missions (at the cost of complexity and emissions).

Importantly, hybrids can be made greener by using Sustainable Aviation Fuel (SAF) or biofuel in the combustion engine. For instance, a small turbine running on a certified SAF could reduce net lifecycle carbon emissions by 50–80% compared to using fossil Jet-A, depending on the SAF's feedstock and production pathway (Tjandra et al., 2020; Undavalli et al., 2023). Thus, a hybrid eVTOL using SAF might achieve significant carbon reduction relative to a conventional helicopter, even though it is not zero-emission. Some industry roadmaps envision initial AAM services utilizing hybrid eVTOLs or even existing light airplanes and helicopters powered by SAF, as a means to "kick-start" the market. Several early AAM pilot programs are likely to rely on retrofitted aircraft; for example, one might use a small hybrid-electric fixed-wing aircraft for regional mobility services in the near term, then transition to a fully electric VTOL in the longer term. Similarly, some of the first "drone taxi" services may utilize existing helicopters running on biofuel to demonstrate the concept of urban air mobility with reduced emissions, until certified eVTOLs become available (Rajendran & Pagel, 2020). Over time, as battery energy density improves and charging networks grow, the need for hybrids may diminish. However, in the 2020s, hybrids represent a practical compromise that extends range and capability while maintaining a smaller environmental footprint compared to traditional aircraft. The

Bell Nexus itself evolved: by 2020, Bell introduced an updated Nexus 4EX concept that could be all-electric for shorter urban missions, with an option to add a hybrid power module for extended range if customers required it (Johnson, 2020). This kind of flexible architecture shows how manufacturers are hedging, designing aircraft that can be primarily electric but with a hybrid boost if needed. From an environmental standpoint, full electrification remains the ideal for zero emissions and minimal noise; yet, hybrid systems can play a role in the transition, ensuring that AAM can meet a variety of use cases, especially longer routes, without waiting for the very latest battery breakthroughs. The key will be to use hybrids economically and smartly, and to supply them with the cleanest possible fuels, so that the overall AAM ecosystem stays on track toward deep decarbonization.

VEHICLE AND INFRASTRUCTURE LIFECYCLE SUSTAINABILITY

To accurately evaluate AAM's actual environmental impact, one must consider not only operational emissions and noise but also the entire lifecycle of both vehicles and infrastructure. This means considering the sustainability of materials and processes used in manufacturing AAM vehicles, the impacts during their use-phase (maintenance, energy consumption, etc.), and what happens at end-of-life (reuse, recycling, or disposal of components). It also encompasses the infrastructure required—vertiports, charging stations, and air traffic management hardware—as well as the footprint of building and maintaining that infrastructure.

Considering the entire lifecycle, the overarching theme is avoiding the transfer of environmental burdens. We do not want AAM to reduce tailpipe emissions but create a mountain of manufacturing emissions or waste. A holistic approach—tracking environmental performance across the supply chain, operational life, and end-of-life—is essential. Encouragingly, many AAM players and regulators are explicitly embracing this. The industry's youth is an advantage: because AAM is still in development, there is an opportunity to "bake in" sustainability from the start, rather than retrofit it later. By planning for recycling, utilizing renewable energy, and sourcing materials responsibly at every stage, AAM operations can align with the principles of a circular economy (Rutkowski, 2021). In doing so, AAM can ensure that its promise of sustainability (e.g., zero-emission flights over our cities) is not undermined by hidden costs in manufacturing or waste disposal. It is not just about *flying green*, but *also about building and retiring green*. Future regulations might formalize this, just as car manufacturers in some regions must comply with end-of-life recycling laws, we may see rules for vertiport sustainability or eVTOL recycling. But even before any mandates, the forward-thinking approach is straightforward: plan AAM as a sustainable system from *cradle to grave*.

AIRSPACE CONGESTION AND ENVIRONMENTAL IMPLICATIONS

When we imagine the skies of the future teeming with air taxis and delivery drones, a concern that naturally arises is airspace congestion. While airspace might seem limitless compared to a packed highway, in practice, aircraft must follow designated routes and maintain separation, so only a specific density of traffic can be handled safely and efficiently in a given volume of airspace. If AAM operations are not well-managed, congestion could lead to suboptimal flight patterns—for example, eVTOLs circling or hovering waiting for a landing slot, which would waste energy and undermine some of the environmental benefits we have discussed. On the flip side, a well-organized AAM traffic management system could optimize routing and scheduling so that flights are direct, minimizing time aloft and energy use.

Airspace congestion management is not just a technical or safety issue; it is an environmental one. By preventing bottlenecks and unnecessary loitering in the air, we can ensure AAM vehicles use energy only to move people and goods to their destinations, thereby avoiding wasted time and power while stuck in aerial "traffic jams." The vision is that AAM flights will be as direct and efficiently timed as possible. Achieving this will require new levels of automation and inter-vehicle communication, which are being actively developed through projects such as NASA's AAM National Campaign and various industry consortia. Early demonstration projects also provide optimism. For example, in flight tests at Pontoise airfield near Paris, CORUS-XUAM partners operated a crewed electric aircraft alongside remotely piloted eVTOL and UAS, coordinating conventional ATM with a digital UTMsystem and successfully exercising rerouting scenarios—including diversions caused by vertiport/airport closure, unexpected FATO occupancy, and two aircraft requiring a shared-path diversion (European Commission, 2022; SESAR Joint Undertaking, 2022; Volocopter, 2022). Similarly, at the Osaka 2025 World Expo, Japan plans to coordinate several eVTOL flights per hour as a showcase of multi-vehicle integration, ensuring each flight takes off and lands on schedule to give spectators a seamless experience (SkyDrive, 2025). These controlled demonstrations are small-scale, but they teach valuable lessons that will inform larger deployments. The bottom line is, if we manage the new "sky highways" well, AAM can deliver its environmental benefits (cleaner air, less noise than current aircraft) without significant new downsides. If we fail to manage it, we may encounter inefficiencies that unnecessarily increase energy use and noise. Fortunately, the consensus in the industry is that innovative air traffic management is an integral part of AAM development, not an afterthought—and this gives hope that as the skies become busier, they will also become *smarter* in tandem.

CASE STUDIES

To illustrate the concepts we've discussed, it's helpful to examine what is happening in various parts of the world at present. Many AAM programs and early implementation projects explicitly highlight environmental considerations, providing real-world examples of how sustainability is being addressed in practice:

United States—NASA's AAM National Campaign and Los Angeles Planning

In the United States, NASA's Advanced Air Mobility National Campaign is a flagship initiative bringing together government, industry, and academia to test

AAM vehicles and operational concepts. A major focus of this campaign is environmental performance, particularly noise. As mentioned earlier, NASA partnered with Joby Aviation to measure the eVTOL's acoustic profile, and the results (around 45 dB(A) in cruise) were impressively low (Blain, 2022). By putting concrete numbers on eVTOL noise, NASA is helping regulators establish noise certification standards that are appropriate for these new aircraft (ICAO, 2022). The FAA, informed by this data, is in the process of drafting eVTOL noise standards to ensure any vehicle entering service meets stringent noise limits (FAA, 2023). On the local level, U.S. cities are also taking initiative. Los Angeles, through its Urban Movement Labs and Department of Transportation, has been proactively planning for urban air mobility with an eye on sustainability and equity. LA conducted a series of community workshops and even set up a "Sound Lab" where residents could listen to simulated eVTOL noise and provide feedback (City of Los Angeles, 2021). This directly influenced the city's policy framework: LA now has a preliminary route network designed to minimize flights over homes, favoring highways and river corridors for aerial routes (City of Los Angeles, 2021). They have also considered environmental justice, aiming to avoid adding noise burden to low-income communities already impacted by other infrastructure. LA's forward-leaning approach serves as a model for how cities can engage the public and incorporate environmental mitigations before the first air taxi takes off. It's a reminder that AAM is as much a local planning challenge as it's a technological challenge.

Europe—Urban-Air Port in the United Kingdom

Europe is seeing significant AAM activity tied to high-profile events and government-supported trials. In the United Kingdom, a notable infrastructure demonstration was the unveiling of the Coventry Air-One vertiport in April 2022. This was the world's first fully functional prototype vertiport, developed by Urban-Air Port with the support of the UK government. The compact vertiport was set up near Coventry's city center for a few weeks as a public demonstration. It hosted drone flights (and was sized for future passenger eVTOLs), and crucially, it operated off-grid with its clean energy system (Sandhu, 2024). The vertiport featured solar panels and a hydrogen fuel cell that charged a battery energy storage system, from which all aircraft charging was conducted, eliminating the need for grid power and making operations essentially emissions-free. This was a powerful proof-of-concept showing that not only the vehicles, but also the ground infrastructure can be made sustainable. Urban-Air Port also emphasized the vertiport's design being modular and temporary—after the demo, it was deconstructed, leaving the site as it was. Following Coventry, the company announced plans for 200 more such vertiports globally, each with a strong focus on renewables and minimal environmental impact. This project won awards for its innovative approach to sustainable infrastructure. It garnered significant public attention, indicating that people are excited about the idea of eco-friendly "mini airports" in cities.

REFERENCES

Acoustic Analytics. (n.d.). Spherical spreading. Retrieved October 27, 2025, from https://www.acousticanalytics.com/modeling/spherical-spreading/

Aggarwal, S., Gupta, P., Mahajan, N., Balaji, S., Singh, K. J., Bhargava, B., & Panda, S. (2023). Implementation of drone-based delivery of medical supplies in North-East India: Experiences, challenges, and adopted strategies. *Frontiers in Public Health, 11*, 1128886. https://doi.org/10.3389/fpubh.2023.1128886

Aminudin, M., Kamarudin, S., Lim, B., Majilan, E., Masdar, M., & Shaari, N. (2022). An overview: Current progress on hydrogen fuel cell vehicles. *International Journal of Hydrogen Energy*. https://doi.org/10.1016/j.ijhydene.2022.10.156

Blain, L. (2022, May 10). NASA acoustic testing puts real numbers on Joby's eVTOL noise signature. *New Atlas*. https://newatlas.com/aircraft/nasa-joby-evtol-noise/

BryceTech. (2023). *Advanced air mobility evidence review* (Independent report prepared for the UK Department for Transport). Department for Transport.

Chatterjee, R. (2025, May 31). How drones are revolutionising air cargo sustainability. *The STAT Trade Times*. https://www.stattimes.com/drones/how-drones-are-revolutionising-air-cargo-sustainability-1355444

Cho, S., & Kim, M. (2022). Assessment of the environmental impact and policy responses for urban air mobility: A case study of the Seoul metropolitan area. *Journal of Cleaner Production, 356*, 132139. https://doi.org/10.1016/j.jclepro.2022.132139

City of Los Angeles. (2021). *Urban air mobility policy framework*. Los Angeles Department of Transportation. https://ladot.lacity.gov/sites/default/files/documents/ladot-uam-policy-framework-considerations.pdf

Discovery of Sound in the Sea. (n.d.). Cylindrical vs. spherical spreading. Retrieved October 27, 2025, from https://dosits.org/science/advanced-topics/cylindrical-vs-spherical-spreading/

Engineering ToolBox. (n.d.). Propagation of sound outdoors – attenuation vs. distance. Retrieved October 27, 2025, from https://www.engineeringtoolbox.com/outdoor-propagation-sound-d_64.html

European Commission, CORDIS. (2022, October 24). *Urban air mobility flight tests in France a success*. CORDIS—EU Research Results.

European Union Aviation Safety Agency. (2021). *Study on the societal acceptance of urban air mobility in Europe*. EASA. easa.europa.eu/sites/default/files/dfu/uam-full-report.pdf

European Union Aviation Safety Agency. (2025, January 8). Type-Certificate Data Sheet for Noise No. EASA.IM.R.120: Robinson R22 (Issue 4). https://www.easa.europa.eu/en/downloads/7979/en

Federal Aviation Administration. (2023). *Innovate28: The Advanced Air Mobility (AAM) implementation plan (Version 1.0)*. FAA. faa.gov/sites/faa.gov/files/AAM-I28-Implementation-Plan.pdf

Flemons, K., Baylis, B., Khan, A. Z., Kirkpatrick, A. W., Whitehead, K., Moeini, S., & Lapointe, S. (2022). The use of drones for the delivery of diagnostic test kits and medical supplies to remote First Nations communities during COVID-19. *American Journal of Infection Control, 50*(8), 849–856. https://doi.org/10.1016/j.ajic.2022.03.004

Friedrich, C., & Robertson, P. (2015). Hybrid-electric propulsion for aircraft. *Journal of Aircraft, 52*(1), 176–189. https://doi.org/10.2514/1.C032660

Gao, Z., Yu, Y., Wei, Q., Topcu, U., & Clarke, J. P. (2024). Noise-aware and equitable urban air traffic management: An optimization approach. *Transportation Research Part C: Emerging Technologies, 165*, 104740. https://doi.org/10.1016/j.trc.2024.104740

Gordo Arias, V. M., Pérez-Castán, J. A., & Sánchez-Medina, L. (2024). Feasibility of conflict prediction of drone trajectories by means of machine-learning techniques. *Aerospace, 11*(12), 1044. https://doi.org/10.3390/aerospace11121044

Hagag, N., & Hoeveler, B. (2025). The feasibility of electric air taxis: Balancing time-savings and CO_2 emissions—A joint case study of respective plans in Paris. *CEAS Aeronautical Journal, 16*(3), 763–788. https://doi.org/10.1007/s13272-025-00811-8

Hermesmann, M., & Müller, T. (2022). Green, turquoise, blue, or grey? Environmentally Friendly hydrogen production in transforming energy systems. *Progress in Energy and Combustion Science*. https://doi.org/10.1016/j.pecs.2022.100996

Huang, Y., Wang, H., & Tzachor, A. (2025). Assessing the environmental impacts associated with China's battery minerals and technologies. *Resources, Conservation and Recycling, 209*, 107978. https://doi.org/10.1016/j.resconrec.2024.107978

International Civil Aviation Organization (ICAO). (2022). Assessment of noise from emerging technology aircraft. In *ICAO environmental report 2022* (Chapter 2). ICAO. https://www.icao.int/sites/default/files/environmental-protection/Documents/EnvironmentalReports/2022/ENVReport2022_Art13.pdf

Johnson, O. (2020, January 6). Bell unveils electric four-ducted Nexus 4EX at CES 2020. *Vertical Magazine*. https://www.verticalmag.com/news/bell-unveils-electric-four-ducted-nexus-4ex-at-ces-2020/

Kang, D., Chen, M., & Ogunseitan, O. (2013). Potential environmental and human health impacts of rechargeable lithium batteries in electronic waste. *Environmental Science & Technology, 47*(10), 5495–5503. https://doi.org/10.1021/es400614y

Kasliwal, A., Furbush, N. J., Gawron, J. H., McBride, J. R., Wallington, T. J., De Kleine, R. D., Kim, H. C., & Keoleian, G. A. (2019). Role of flying cars in sustainable mobility. *Nature Communications, 10*, 1555. https://doi.org/10.1038/s41467-019-09426-0

Kaya, D., Kılıç, F. Ç., & Öztürk, H. H. (2021). Energy efficiency in electric motors. In D. Kaya, F. Ç. Kılıç, & H. H. Öztürk (Eds.), *Energy management and energy efficiency in industry* (Green Energy and Technology). Springer. https://doi.org/10.1007/978-3-030-25995-2_12

Khavarian, K., & Kockelman, K. M. (2023). Life-cycle analysis of electric vertical take-off and landing vehicles. *Transportation Planning and Technology, 47* (8), 1227–1242. https://doi.org/10.1080/03081060.2023.2217163

Leon, E., & Miller, S. A. (2020). An applied analysis of the recyclability of electric vehicle battery packs. *Resources, Conservation and Recycling, 159*, 104593. https://doi.org/10.1016/j.resconrec.2019.104593

Lewis-Kraus, G. (2024, April 15). *Are flying cars finally here? The New Yorker.*

Li, J., Lu, Y., Yang, T., Ge, D., Wood, D. L., & Li, Z. (2020). Water-based electrode manufacturing and direct recycling of lithium-ion battery electrodes—A green and sustainable manufacturing system. *iScience, 23*(12), 101081. https://doi.org/10.1016/j.isci.2020.101081

Li, X., Zhang, Y., & Chen, L. (2024). Advancing social equity in urban UAV logistics: Insights from academic literature and social media. *Drones, 8*(11), 688. https://doi.org/10.3390/drones8110688

Liu, M., Hao, H., Geng, Y., Dou, H., Sun, X., Bai, F., Qu, X., Wang, K., Liu, Z., & Zhao, F. (2025). Adopting electric vertical takeoff and landing aircraft for green transport transformation. *Fundamental Research*. https://doi.org/10.1016/j.fmre.2025.04.002

Marksel, M., & Brdnik, P. (2023). Comparative analysis of direct operating costs: Conventional vs. hydrogen fuel cell 19-seat aircraft. *Sustainability*. https://doi.org/10.3390/su151411271

Orikpete, O. F., Leton, T. G., Amah, V. E., & Ewim, D. R. E. (2020). An assessment of the impact of helicopter noise: Case study of Mgbuoshimini community, *Nigeria. Journal of Earth and Environmental Science Research*, 2(2), 1–16.

Pellow, M., Emmott, C., Barnhart, C., & Benson, S. (2015). Hydrogen or batteries for grid storage? A net energy analysis. *Energy and Environmental Science*, 8, 1938–1952. https://doi.org/10.1039/C4EE04041D

PwC. (2022). *Skies without Limits v2.0: A refreshed look at how drones could impact the UK's economy, jobs, productivity and quality of life* (July 2022). PwC UK.

Rajendran, S., & Pagel, E. (2020). Recommendations for emerging air taxi network operations based on online review analysis of helicopter services. *Heliyon*, *6*, e05581. https://doi.org/10.1016/j.heliyon.2020.e05581

Rettenmeier, M., Möller, M., & Sauer, A. (2024). Disassembly technologies of end-of-life automotive battery packs as the cornerstone for a circular battery value chain: A process-oriented analysis. *Resources, Conservation and Recycling*, *209*, 107786. https://doi.org/10.1016/j.resconrec.2024.107786

Rizzi, S. A. (2022, September 21–22). Characterization of Urban Air Mobility Vehicle Operational Noise and Community Noise Impact [Conference presentation]. Académie de l'Air et de l'Espace (AAE). https://ntrs.nasa.gov/api/citations/20220013831/downloads/220922%20Urban%20Transport%20of%20Passengers%20by%20eVTOL%20-%20Rizzi_v2c.pdf

Rizzi, S. A., Cheng, R., & Page, J. (2023). Prediction of noise–power–distance data for urban air mobility vehicles. NASA Technical Reports Server. https://ntrs.nasa.gov/api/citations/20230000171/downloads/rizzi%20et%20al%202023predictionofnoise.pdf

Rizzi, S. A., Huff, D. L., Boyd, D. D. Jr., Bent, P., Henderson, B. S., Pascioni, K. A., Sargent, D. C., Josephson, D. L., Marsan, M., He, H., & Snider, R. (2020). Urban air mobility noise: Current practice, gaps, and recommendations (NASA/TP–2020-5007433). NASA Technical Reports Server. https://ntrs.nasa.gov/api/citations/20205007433/downloads/NASA-TP-2020-5007433.pdf

Rodrigues, T., Patrikar, J., Oliveira, N., Matthews, H. S., Scherer, S., & Samaras, C. (2022). Drone flight data reveal energy and greenhouse gas emissions savings for very small package delivery. *Patterns*, *3*, 100569. https://doi.org/10.1016/j.patter.2022.100569

Rutkowski, M. (2021). The areas of impact of the circular economy in passenger air transport. In C. Bode (Ed.), *Intelligent solutions for cities and mobility of the future* (Lecture Notes in Networks and Systems). Springer. https://doi.org/10.1007/978-3-030-91156-0_10

Sandhu, R. (Guest). (2024, August 1). How Urban-Air Port is revolutionising infrastructure for advanced air mobility [Audio podcast episode]. In *Sustainability in the air*. SimpliFlying. https://www.buzzsprout.com/1976041/episodes/15512574

Sarkar, M., Yan, X., Girma, A., & Homaifar, A. (2023). A comprehensive eVTOL performance evaluation framework in urban air mobility. In K. Arai (Ed.), *Intelligent Systems and Applications (IntelliSys 2022) (Lecture Notes in Networks and Systems*, Vol. 542, pp. 459–474). Springer, Cham. https://doi.org/10.1007/978-3-031-16072-1_34

Schäffer, B., Pieren, R., Heutschi, K., Wunderli, J. M., & Becker, S. (2021). Drone noise emission characteristics and noise effects on humans—A systematic review. *International Journal of Environmental Research and Public Health*, *18*(11), 5940. https://doi.org/10.3390/ijerph18115940

Schiller, N. (2023). *Advanced air mobility: Environmental impact (Noise and emissions)* [Conference presentation]. NASA Langley Research Center. https://ntrs.nasa.gov/citations/20230012372

Schiller, N. (2023, September 11). *Advanced air mobility: Environmental impact (noise and emissions)* [Presentation]. NASA Technical Reports Server. https://ntrs.nasa.gov/api/citations/20230012372/downloads/AAM%20Environmental%20Impact_v2.pdf

SESAR Joint Undertaking. (2022). *CORUS-XUAM partners complete deconfliction flight tests in France.*

SkyDrive. (2025, August 28). *SkyDrive releases onboard footage of demonstration flights at Expo 2025 Osaka* [Press release]. https://en.skydrive2020.com/archives/16292

Spindlegger, A., Slotyuk, L., Jandric, A., De Souza, R. G., Prenner, S., & Part, F. (2025). Environmental performance of second-life lithium-ion batteries repurposed from electric vehicles for household storage systems. *Sustainable Production and Consumption*, *54*, 227–240. https://doi.org/10.1016/j.spc.2025.01.003

Sripad, S., & Viswanathan, V. (2021). The promise of energy-efficient battery-powered urban aircraft. *Proceedings of the National Academy of Sciences*, *118*(45), e2111164118. https://doi.org/10.1073/pnas.2111164118

Swartz, K. I., & Hirschberg, M. (2020, March 5). *The first electric VTOL unicorn: Joby Aviation. Vertiflite* (Vertical Flight Society), Mar/Apr 2020.

Technical University of Munich. (2020, June 17). *Valuation of start-up Lilium tops $1 billion* [Press release].

Tian, J., & Chen, M. (2014). Sustainable design for automotive products: Dismantling and recycling of end-of-life vehicles. *Waste Management*, *34*(2), 458–467. https://doi.org/10.1016/j.wasman.2013.11.005

Tjandra, A., Routhieau, V., Chishty, W., Basset, P., Peluso, R., & Bérat, C. (2020). Greenhouse gas (GHG) reduction study for the rotorcraft industry. In A. Datta (Chair), *Proceedings of the Vertical Flight Society 76th Annual Forum*. Vertical Flight Society.

Tjandra, A., Basset, P.-M., Routhieau, V., Chishty, W., Bérat, C., & Peluso, R. (2020). *Green House Gas (GHG) reduction study for the rotorcraft industry*. In *Proceedings of the 76th Annual Forum* (Virtual, October 5–8, 2020). The Vertical Flight Society.

Undavalli, V., Olatunde, O. B. G., Boylu, R., Wei, C., Haeker, J., Hamilton, J., & Khandelwal, B. (2023). Recent advancements in sustainable aviation fuels. *Progress in Aerospace Sciences*, *136*, 100876. https://doi.org/10.1016/j.paerosci.2022.100876

Velaz-Acera, N., Ruiz-Marín, R., & Borge-Diez, D. (2025). Comparative economic, energy, and environmental analysis of fuel cell and electric eVTOL systems: Case study of Iberian Peninsula. *Journal of Cleaner Production*, *495*, 145027. https://doi.org/10.1016/j.jclepro.2025.145027

Volocopter. (2022). *M3 Systems, Pipistrel, and Volocopter complete deconfliction flight tests in France.*

Weitering, H. (2024, March 15). Alaka'i makes the case for hydrogen-powered eVTOL aircraft. *Aviation International News*. https://www.ainonline.com/aviation-news/futureflight/2024-03-15/alakai-makes-case-hydrogen-powered-evtol-aircraft

Yoo, B., Wilailak, S., Bae, S., Gye, H., & Lee, C. (2021). Comparative risk assessment of liquefied and gaseous hydrogen refueling stations. *International Journal of Hydrogen Energy*. https://doi.org/10.1016/j.ijhydene.2021.08.073

Zart, N. (2024, March 26). NREL says more energy and storage needed for eVTOL charging. *Electric Air Mobility News*. https://www.electricairmobility.news/2024/03/26/nrel-study-aam-evtol-energy-storage-grid-requirement/

Zhang, J., Roumeliotis, I., & Zolotas, A. (2022). Model-based fully coupled propulsion-aerodynamics optimization for hybrid electric aircraft energy management strategy. *Energy*, *245*, 123239. https://doi.org/10.1016/j.energy.2022.123239

Zorpette, G., & Ackerman, E. (2022, February 22; updated 2024, March 27). *EVTOL companies are worth billions—Who are the key players? IEEE Spectrum.*

9 Societal Dimensions

INTRODUCTION

In December 2016, at Kabgayi District Hospital in Rwanda, a two-year-old girl on the brink of death from severe malaria was saved by an unlikely hero: a drone. The fixed-wing vehicle buzzed overhead and released a red package containing urgently needed blood (see Figure 9.1)—a trip that would have taken hours by road but arrived in minutes (Fleming, 2018). Moments later, the child opened her eyes, becoming the first person in the world whose life was documented as being saved by an autonomous medical drone delivery. The story demonstrates how advanced air mobility (AAM) is already closing critical logistics gaps and saving lives.

Across the globe, a very different AAM scenario is unfolding. In late 2024, as captured in Figure 9.2, spectators watched a Joby Aviation electric air-taxi prototype lift off at a Toyota test facility, silently skimming past Mount Fuji (Joby Aviation, 2024). The all-electric, vertical-take-off-and-landing (eVTOL) aircraft—bankrolled in part by Toyota—demonstrated how passengers might soon hop over gridlock in quiet flying taxis. The public flight showcased the vehicle's technical maturity and hinted at a future where urban air mobility becomes a daily reality.

These snapshots highlight AAM's promise to make transport more efficient, accessible, and even lifesaving, yet they also hint at the challenges ahead. In suburban Australia, for instance, residents protested when delivery drones passed overhead dozens of times a day, complaining the buzz sounded like "*a swarm of mosquitoes*" and insisting *it's gone too far* (Stonor, 2022). Such community feedback highlights a crucial point: AAM's success depends not only on engineering but also on public acceptance and thoughtful social integration (Kloss & Riedel, 2021).

In other words, AAM is as much a social revolution as a technological one. Even the best-engineered drone or air taxi will fail if people do not trust it, feel safe, and see tangible benefits. To embed flying taxis and drones sustainably into everyday transport, we address all the relevant interlocking societal dimensions in the chapter.

PUBLIC ACCEPTANCE AND PERCEPTION OF SAFETY

Public acceptance is widely recognized as a critical enabler for AAM's success. Surveys consistently show that the general public's willingness to embrace flying taxis and delivery drones hinges first and foremost on safety. In a broad 2021 survey spanning the United States, Europe, India, Brazil, and China, over 60% of respondents identified safety as their top concern about passenger AAM vehicles (Kloss & Riedel, 2021). Safety outweighed other issues by a large margin—for example, concerns about price ranked a distant second, cited by only around 10–20% of respondents. Likewise, a European Union study by EASA found that citizens expect AAM operations to be as safe as current commercial aviation, reflecting the aviation industry's strong safety record and the public's low

DOI: 10.1201/9781003546306-12

FIGURE 9.1 A Rwandan healthcare worker holds a blood delivery package dropped by a Zipline drone, with the parachute still attached. (Photo: Jason Florio/Zipline; Ackerman, 2019.)

FIGURE 9.2 A prototype Joby Aviation air taxi flies in front of Mount Fuji during a 2024 demonstration. (Becker, 2024)

tolerance for risk in new transport modes (EASA, 2021; Pritchard, 2021). Public concern is especially pronounced regarding autonomous flight—many people are uneasy about the idea of pilotless air taxis until the technology proves itself reliable (Kloss & Riedel, 2021). Even with human pilots, caution exists: some consumers admit hesitancy about flying in any small aircraft, which translates into wariness of AAM until safety is demonstrated at scale (Al Haddad et al., 2020; Kloss & Riedel, 2021).

Initial attitude surveys indicate that many people are cautiously optimistic about AAM, provided safety and other concerns are effectively addressed. In Europe, EASA's landmark 2021 study across six urban regions found 83% of respondents held a positive initial attitude toward urban air mobility, and 71% said they were ready to try UAM services in principle (EASA, 2021). However, this openness comes with clear conditions: when asked about specific worries, participants in that study flagged *safety, noise, and environmental impacts* as top concerns for both air taxis and drones (EASA, 2021). Notably, about 38% of European respondents cited safety as a concern with passenger air taxis (similar to the share citing noise or environmental effects), and 44% cited safety as their highest concern with delivery drones (Pritchard, 2021). Noise plays a dual role in acceptance—not only as a potential nuisance, but also as a proxy for safety in the public's mind. Unfamiliar or loud noises from low-flying vehicles can alarm people and erode trust in the authorities. In a community survey related to unmanned traffic management, over 55% of respondents identified aircraft safety as the most important factor for acceptance, followed closely by the *type of noise* (49%) and *volume of noise* (48%) generated by AAM vehicles (Vertx Partners, 2024). The timing and altitude of flights also mattered to nearly half of those surveyed (Vertx Partners, 2024), indicating that the timing and height at which AAM vehicles fly over communities will affect public comfort. Furthermore, emerging research on noise perception suggests that drone sounds are inherently more annoying to people than traditional vehicle noise at the same decibel levels, due to their distinct pitch and character (Kawai et al., 2024). This highlights why controlling noise is so crucial for public acceptance.

Despite these concerns, there are promising signs of emerging public support for AAM. In the Airbus UTM survey mentioned above, nearly 45% of respondents expressed support or firm support for AAM implementation in their area, and approximately 41% even deemed AAM "safe" or "very safe" at this early stage of the industry (Vertx Partners, 2024). Public sentiment is likely to evolve as people become more familiar with the technology through demonstrations and media exposure. Demographics play a notable role in acceptance: younger adults tend to be more open to AAM, whereas older individuals voice greater safety concerns (Al Haddad et al., 2020; Kloss & Riedel, 2021). For instance, willingness to try an air taxi or drone delivery drops significantly with age—a trend observed in both the EASA and McKinsey surveys (EASA, 2021; Kloss & Riedel, 2021). This suggests that targeted outreach may be needed to address the specific worries of older demographics. On the other hand, experienced tech adopters and frequent travelers—such as current users of ride-hailing services or premium transportation—exhibit a higher willingness to be early adopters of AAM (Kloss & Riedel, 2021). Familiarity with innovative transport modes and the lure of time-saving convenience were cited as reasons these groups would consider switching to air taxis.

Crucially, use cases and perceived public benefit strongly influence acceptance. Surveys show that people are most receptive to AAM when it serves precise societal needs such as emergency response, medical transport, or public safety. In the EASA study, the top anticipated use cases were emergency-related: for example, airlifting an injured person to a hospital or delivering medical supplies ranked highest in public support (over 40% of respondents selected these as desirable services) (EASA, 2021; Pritchard, 2021). In fact, improved emergency response times were seen as the number one expected benefit of urban air mobility (around 71% agreement), even above reducing traffic congestion or pollution (Pritchard, 2021). The implication is that AAM will gain trust faster if it is introduced first in socially valuable roles (medical evacuation, rescue, firefighting, disaster relief) rather than purely as a premium taxi for the affluent. Early demonstration programs have taken this to heart; for example, in Japan and the United States, pilot projects are testing the delivery of blood and vaccines via medical drones to remote clinics. Communities have broadly welcomed these trials because they save lives in areas hard to reach by road (Fleming, 2018). In Rwanda, drone delivery services now transport over 35% of all blood supplies outside the capital city, dramatically reducing delivery times to rural hospitals and garnering broad public support due to the technology's life-saving impact (Fleming, 2018). These kinds of use cases cast AAM in a favorable light and build a narrative of public good around the technology.

Such an impact on the local community has been established by Zipline's medical-drone service, which, since launching in 2016, has hit several cumulative delivery milestones, as illustrated in Figure 9.3. In its first year, it completed roughly 7,000 flights in Rwanda, with the first documented life saved at Kabgayi District Hospital. By 2019, after expanding to Ghana, Zipline surpassed 100,000 flights. The company marked 200,000 in 2021 on its five-year anniversary, crossed 500,000 in 2023, and on April 19, 2024, announced its one-millionth medical delivery.

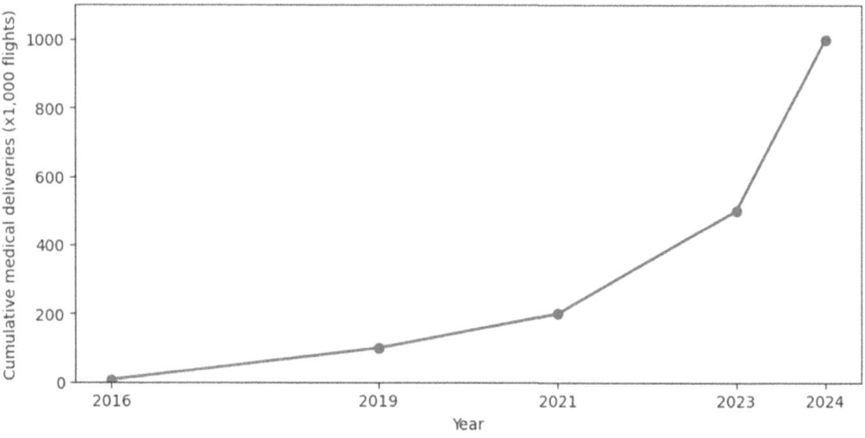

FIGURE 9.3 Lives saved by medical drones (2016–2024). (Credit: Author, Data source: Baker, 2018; Citi Newsroom, 2021; Johnson, 2023; Kolodny, 2019; Miller, 2023; Zipline, n.d., 2024.)

Building public trust in AAM will require proactive strategies that extend beyond simply maintaining good safety records. Effective communication and community engagement are essential. Regulatory bodies and industry groups have begun public outreach campaigns to educate communities about AAM's benefits and to address concerns transparently. For instance, NASA has been studying public priorities for AAM to ensure that *public good* use cases are emphasized in early deployments—part of a deliberate effort to enhance trust and demonstrate value (Reed, 2023). Hands-on demonstrations can also help demystify AAM. In one example, West Virginia's AAM integration initiative included a mobile simulator of drone air traffic control that toured local events, allowing residents to interact with the technology (Vertx Partners, 2024). Such experiences helped dispel misconceptions and increase people's comfort by making AAM tangible rather than abstract. Experts often emphasize the importance of securing a "social license" for AAM—meaning that communities grant informal permission for operations because they feel included in the process and see clear benefits. As one industry advocacy group put it, before eVTOL air taxis fly routinely in our cities, they will need not just an aviation certificate but also essentially a license from the community they serve (Aerobility & CIVATAglobal, 2021; Aerobility & CIVATAglobal, 2021). Earning this social license involves early engagement with local stakeholders, sharing safety and noise data openly, addressing environmental and privacy concerns, and introducing new services gradually so the public can adapt. In summary, public acceptance of AAM is achievable, but it hinges on demonstrating safety, minimizing intrusions such as noise, showing clear societal benefits, and actively involving the public in the introduction of this new technology. Cautious optimism can be converted into confidence if industry and regulators put people first in AAM's rollout.

EQUITY AND ACCESS IN ADVANCED AIR MOBILITY

A critical question for the sustainability of AAM is: Who will benefit from these new air mobility services? Without deliberate planning, AAM could easily become an exclusive domain of wealthy early adopters, providing fast aerial commutes for a few while bypassing the needs of many. History shows that new transport technologies often benefit the elite first before eventually "trickling down." They can also impose negative impacts on less advantaged communities if not managed with equity in mind (Reed, 2023). To avoid exacerbating existing inequities, stakeholders are now making equitable access a core principle in the deployment of AAM.

GEOGRAPHIC AND ECONOMIC EQUITY

Initial AAM services are likely to target lucrative, high-demand routes—for example, connecting city centers to airports or linking affluent suburbs and business districts, where customers are willing to pay a premium for the time saved. Indeed, early business plans (such as those by Uber Elevate and various eVTOL startups) emphasize airport shuttles and downtown air-taxi corridors. If left purely to market forces, such a rollout could create a two-tier transportation system where the wealthy literally fly over congested ground traffic, while lower-income commuters remain stuck on buses and highways below. Market-driven deployment alone would favor areas with high-paying demand, potentially leaving "transit deserts" and underserved neighborhoods even further behind. To counter this, policy interventions and inclusive business models are being explored to make AAM more accessible. For example, some city planners are considering requirements that a certain percentage of vertiports, or routes, must serve underserved areas, or that pricing should include affordable options (akin to transit subsidies or tiered fares) (McNab, 2024; Reed, 2023). In the United States, a few states are even introducing legislation to embed equity into AAM planning. In California, proposed bills on AAM infrastructure have explicitly mandated that planning processes consider equitable access and fair distribution of benefits as key criteria (Reed, 2023). The goal of these measures is to avoid repeating the mistakes of past transportation innovations (such as urban freeways that often divided or burdened low-income communities) by ensuring AAM does not just cater to the privileged.

One approach to advance equity is to prioritize public-oriented AAM use cases in early phases. For instance, cargo drones for medical deliveries can serve rural villages and low-income regions, improving access to healthcare and essential goods. Such services are already active in countries like Rwanda and Ghana, where they target communities lacking reliable ground logistics and have been met with enthusiasm (Fleming, 2018). In developed nations, public-private partnerships are being envisioned—for example, utilizing eVTOL aircraft for emergency response in hard-to-reach areas, or providing first- and last-mile air links for remote towns. By demonstrating value for diverse communities early on, AAM can build broader political and public support beyond just wealthy users.

Additionally, pricing strategies will affect economic equity. Initially, AAM trips are expected to be expensive—early estimates suggest that they will be several times the cost of an equivalent car ride or taxi for the same distance (Kloss & Riedel, 2021). Over time, economies of scale and competition could drive prices down. Regulators and cities might encourage pooled air-taxi rides (shared trips) or integrate AAM flights into public transit subscription systems to make them more affordable for middle-class and working-class users. Ensuring robust competition among multiple AAM operators and vertiport providers will also be crucial; it can prevent monopolies and foster price reductions while maintaining service quality (Reed, 2023). The underlying principle is that AAM should eventually function not just as a luxury service, but as a practical transportation option that a broad segment of the population can utilize.

Physical Accessibility

Equity is not just about economics; it is also about physical ability and inclusive design. AAM vehicles and infrastructure must be designed to accommodate people with disabilities and mobility limitations from the outset. Advocates point out that those who could benefit the most from on-demand aerial mobility include individuals for whom traditional transit is inconvenient or inaccessible, such as persons with disabilities or the elderly who cannot easily use buses, trains, or cars (Aerobility & CIVATAglobal, 2021). If designed correctly, eVTOL air taxis could provide point-to-point transportation for people with reduced mobility, but only if accessibility features are incorporated early in the design process. This translates to considering wheelchair access at vertiports and on aircraft, as well as providing tactile and audio aids for visually impaired travelers and implementing boarding procedures that accommodate those with limited mobility. A 2021 white paper by the U.K. disability flying charity Aerobility and the urban air transport association CIVATAglobal urged the AAM industry to involve disabled-user groups "at the earliest possible stage" of design and planning (Aerobility & CIVATAglobal, 2021). The argument is that retrofitting accessibility later would be more difficult and costly, whereas considering it now can bake inclusivity into the DNA of AAM systems. For example, aviation regulators may need to adapt certification standards to account for the safe carriage of wheelchairs or service animals on eVTOL flights, as well as to ensure that emergency procedures cover passengers with disabilities. Ground facilities (vertiports) should similarly adhere to universal design principles, such as ADA-compliant standards in the United States (Hums et al., 2016), incorporating features like elevators or ramps in addition to stairs, clear signage, and staff training to assist passengers with special needs. By doing so, AAM can avoid inadvertently excluding a significant portion of the population and instead become a new mobility option that *enhances* independence for people with disabilities and the elderly. In short, inclusive design from the outset leads to a more sustainable and widely adopted AAM system.

WORKFORCE IMPACTS: JOB CREATION AND DISPLACEMENT

The advent of AAM will not only transform how people and goods move—it will also have significant effects on the workforce and jobs. As a high-tech emerging industry, AAM will be able to create entirely new categories of employment, demand new skill sets, and potentially disrupt some existing occupations in aviation and transportation.

On the optimistic side, numerous studies project that AAM will be a significant job creator over the coming decades. For example, a report by the Aerospace Industries Association (AIA) and Deloitte estimated that by 2035, the AAM sector could directly employ over 280,000 people in the United States (Hussain & Silver, 2021). This figure—roughly 8% of the current U.S. aerospace and defense workforce—includes jobs in vehicle manufacturing, maintenance, flight operations, infrastructure construction, air traffic management, and supporting services. The majority of those jobs (around 234,000) would serve domestic AAM operations, with the rest tied to export markets as U.S. firms build vehicles and systems for global markets.

Globally, similar growth is anticipated. Countries positioning themselves as AAM leaders foresee tens of thousands of new high-tech and aviation jobs as the market matures (McNab, 2024). Regions that invest early in AAM research and pilot programs aim to become industry hubs and capture these jobs. For example, states like Ohio and West Virginia in the United States are actively courting the AAM industry as a source of economic revitalization in communities affected by industrial decline (Vertx Partners, 2024).

Categories of new jobs emerging with AAM could include, but not be limited to, aircraft design and manufacturing engineers, infrastructure engineers and construction workers, pilots and remote operators, maintenance technicians, air traffic and fleet management personnel, vertiport operations and customer service staff, data analysts and software developers, and regulatory and planning professionals.

Many of these roles are high-skill positions that are likely to offer good wages, particularly in engineering and software fields. Industry studies emphasize that AAM can create high-quality jobs and that proactive workforce development is necessary to fill them (Hussain & Silver, 2021). In the Deloitte/AIA analysis, the authors note the opportunity to develop talent pipelines for aerospace careers and to inspire the next generation of engineers and pilots through the excitement of AAM. However, realizing these job gains will require significant investment in training and education. As one industry panelist put it, the sector's growth "requires thoughtful planning and infrastructure development to ensure safety… [and] the enormous potential of the sector to create jobs and opportunities" (Reed, 2023).

Initiatives are already underway to upskill workers from adjacent industries. For example, programs to retrain automotive manufacturing workers in electric aircraft production, or to help helicopter mechanics learn to service eVTOLs, can smooth the labor transition. Some universities and aviation training providers have begun offering dedicated AAM/UAM education, ranging from specialized centers and workforce-development programs to formal postgraduate degrees focused on advanced air mobility operations and management (Delaware State University CAAM, n.d.; University of West London, n.d.; ITAérea, n.d.). These efforts aim to avoid a talent bottleneck in the AAM industry and also to allow workers from declining sectors (like traditional aviation manufacturing or ground transportation) to pivot into AAM roles.

At the same time, it is important to consider the potential job displacement or disruption that AAM could cause in existing sectors. AAM will partly overlap or compete with some current jobs in aviation and logistics. However, the net impact is expected to be positive (more jobs created than lost) since AAM is opening new markets rather than just automating old ones (McNab, 2024). Key areas that could be affected include helicopter pilots and crews, ground transportation operators, airport and airline personnel, as well as logistics and warehousing personnel.

REFERENCES

Ackerman, E. (2019, October 14). *In the air with Zipline's medical delivery drones*. IEEE Spectrum. https://spectrum.ieee.org/in-the-air-with-ziplines-medical-delivery-drones

Aerobility & CIVATAglobal. (2021, January 20). Urban air mobility must incorporate the needs of disabled citizens [Press release]. *sUAS News*. https://www.suasnews.com/2021/01/urban-air-mobility-must-incorporate-the-needs-of-disabled-citizens/

Al Haddad, C., Chaniotakis, E., Straubinger, A., Plötner, K., & Antoniou, C. (2020). Factors affecting the adoption and use of urban air mobility. *Transportation Research Part A: Policy and Practice*, *132*, 696–712. https://doi.org/10.1016/j.tra.2019.12.020

Baker, A. (2018, May 31). Zipline's drones are saving lives. *Time*. https://time.com/5295821/ziplines-drones-are-saving-lives/

Becker, F. (2024, November 4). *Joby Aviation completes first air taxi flight in Japan with Toyota [Video]*. Electrek. https://electrek.co/2024/11/04/joby-aviation-completes-first-air-taxi-flight-japan-toyota-video/

Citi Newsroom. (2021, October 29). Zipline celebrates 5 years in flight, surpasses 200,000 deliveries. https://citinewsroom.com/2021/10/zipline-celebrates-5-years-in-flight-surpasses-200000-deliveries/

Delaware State University, College of Business. (n.d.). *Center for Advanced Air Mobility*.

European Union Aviation Safety Agency. (2021). *Study on the societal acceptance of urban air mobility in Europe*. https://www.easa.europa.eu/sites/default/files/dfu/uam-full-report.pdf

Fleming, S. (2018, December 11). In Rwanda, high-speed drones are delivering blood to remote communities. *World Economic Forum*. https://www.weforum.org/stories/2018/12/in-rwanda-drones-are-delivering-blood-to-remote-communities/

Hums, M. A., Schmidt, S. H., Novak, A., & Wolff, E. A. (2016). Universal design: Moving the Americans with Disabilities Act from access to inclusion. *Journal of Legal Aspects of Sport*, *26*(1), 36–51. https://doi.org/10.1123/jlas.2015-0011

Hussain, A., & Silver, D. (2021, January 26). Advanced air mobility: Can the United States afford to lose the race? *Deloitte Insights*. https://www.deloitte.com/us/en/insights/industry/aerospace-defense/advanced-air-mobility.html

ITAérea Aeronautical Business School. (n.d.). *Master in Advanced Air Mobility*.

Joby Aviation, Inc. (2024, November 4). Toyota and Joby complete first air taxi flight in Japan [Press release]. https://ir.jobyaviation.com/news-events/press-releases/detail/113/toyota-and-joby-complete-first-air-taxi-flight-in-japan

Johnson, W. (2023, February 2). Zipline reaches 500,000 automated drone delivery milestone. *Teslarati*. https://www.teslarati.com/zipline-500k-automated-drone-delivery-milestone/

Kawai, C., Jäggi, J., Georgiou, F., Meister, J., Pieren, R., & Schäffer, B. (2024). Short-term noise annoyance towards drones and other transportation noise sources: A laboratory study. *The Journal of the Acoustical Society of America*, *156*(4), 2578–2595. https://doi.org/10.1121/10.0032386

Kloss, B., & Riedel, R. (2021, June 1). Up in the air: How do consumers view advanced air mobility? McKinsey & Company. https://www.mckinsey.com/industries/aerospace-and-defense/our-insights/up-in-the-air-how-do-consumers-view-advanced-air-mobility

Kolodny, L. (2019, May 17). Zipline, which launched in Rwanda, now a unicorn. *CNBC Africa*. https://www.cnbcafrica.com/2019/zipline-which-in-rwanda-now-a-unicorn

McNab, R. M. (2024). Advanced air mobility, economic impacts, and equity considerations. *Journal of Economic Analysis*, *3*(2), 61–79.

Miller, B. (2023, December 20). Measuring the impact of drone deliveries. *UDaily* (University of Delaware). https://www.udel.edu/udaily/2023/december/research-robotics-zipline-drones-deliveries-olga-lebed/

Pritchard, J. (2021, May 20). EASA reveals results of first study on public acceptance of UAM in European cities. *eVTOL Insights*. https://evtolinsights.com/easa-reveals-results-of-first-study-conducted-on-public-acceptance-of-urban-air-mobility-in-european-cities/

Reed, J. (2023, August 3). Industry experts weigh in on equity, safety, and efficiency in advanced air mobility. *Avionics International*. https://www.aviationtoday.com/2023/08/03/industry-experts-weigh-in-on-equity-safety-and-efficiency-in-advanced-air-mobility/

Stonor, C. (2022, September 28). Wing drone delivery noise: Australians complain, "It's gone too far!". *eVTOL Insights*. https://evtolinsights.com/wing-drone-delivery-noise-australians-complain-its-gone-too-far/

University of West London. (n.d.). *Advanced Air Mobility Management (MSc)*.

Vertx Partners. (2024, June 6). Building public trust in advanced air mobility. https://vertxpartners.org/building-public-trust-in-aam/

Zipline. (n.d.). Zipline fact sheet. Retrieved October 27, 2025, from https://www.zipline.com/about/zipline-fact-sheet

Zipline. (2024, April 19). Zipline makes one millionth commercial autonomous drone delivery and expands U.S. deliveries to include Panera Bread, Memorial Hermann Health System, and Jet's Pizza. https://www.zipline.com/newsroom/news/announcements/zipline-makes-one-millionth-commercial-autonomous-drone-delivery-and-expands-u-s-deliveries-to-include-panera-bread-memorial-hermann-health-system-and-jet-s-pizza

10 Financial Viability

INTRODUCTION

On a humid August morning in 2023, a sleek white aircraft with a 50 ft wingspan lifted off from Burlington, Vermont, on a journey few electric aircraft had ever attempted. Beta Technologies' ALIA (Figure 10.1)—an eVTOL designed for cargo and, eventually, passenger missions—flew more than 1,500 nautical miles in nine legs, recharging at small airports along the U.S. East Coast before touching down at Eglin Air Force Base in Florida (Finnerty, 2023). Far from a publicity stunt, the flight was part of the U.S. Air Force's *Agility Prime* program, which pays start-ups for real-world flight hours and data. For Beta, the cross-country trek validated its battery-swap concept and, more importantly, triggered milestone payments that extended its cash runway. For the Air Force, it provided performance data on an aircraft that might one day move supplies or personnel at a fraction of today's operating costs.

Stories like Beta's capture the core tension of advanced air mobility (AAM): stunning technical progress paired with ruthless financial realities. Certification campaigns can consume nearly $1 billion before the first paying passenger boards (Deloitte, 2023, p. 4). Timelines slip, capital markets tighten, and even well-funded pioneers such as Lilium and Volocopter have flirted with insolvency (Villamizar, 2025). Yet investors continue to write checks because credible forecasts still place the global eVTOL/UAM total addressable market at about $1 trillion by 2040 (base case)—and extend to around $9 trillion by 2050 under assumptions that include major advances in autonomy and enabling business models (Jonas et al., 2021).

Yet there is one more milestone that completes the picture: in August 2022, United Airlines made a $10 million pre-delivery payment (cash deposit) to Archer Aviation associated with 100 of Archer's initial production eVTOL aircraft, the Midnight (Archer Aviation, 2022). In traditional aerospace, advance payments are routine; in a nascent sector, they are a vote of confidence and a discounted source of working capital. United is not alone, as Japan Airlines, American, and several leasing giants have placed provisional or firm orders, hoping to turn time savings into a competitive advantage on the ground as much as in the sky (Brinkmann, 2024). Each deal signals to regulators, lawmakers, and financiers that AAM is evolving from a concept into a commercial asset class.

For deeper insight, Figure 10.2 illustrates that funding into the U.S. AAM ecosystem is a key enabler of airframe orders, and therefore of industry growth. Similarly to the airplane industry, moving from prototype to certification and scale requires sustained, high-magnitude capital, so order momentum tends to strengthen in high-investment periods and soften when cash-crunch conditions restrict financing (Brunelli et al., 2023; National Aviation Authorities Network [NAA Network], 2025; National Renewable Energy Laboratory [NREL], 2023; Reuters, 2024; Straubinger et al., 2021; U.S. Government Accountability Office [GAO], 2022).

DOI: 10.1201/9781003546306-13

FIGURE 10.1 Beta Technologies' ALIA VTOL. (Credit: BETA Technologies.)

Against this evolving environment, this chapter asks one question: *Can the math keep up the hype?* The answer will be examined through a pragmatic lens on AAM finance: optimistic yet rigorous, as achieving resilient sustainability in AAM is not only about engineering robust aircraft and ensuring safe operations; it is equally

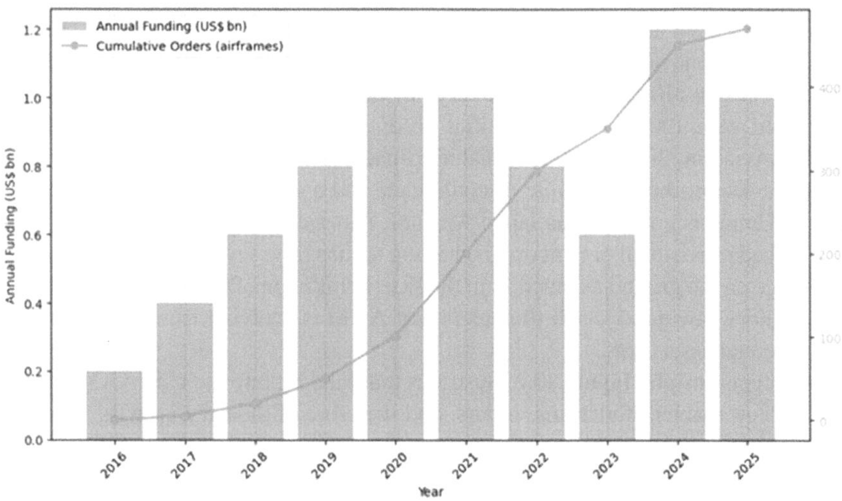

FIGURE 10.2 Capital Inflow vs AAM Airframe Orders Timeline in USA (2016–2025). (Credit: Author, Data source: Brinkmann, 2022; Ganapavaram & Tiwary, 2024; Archer Aviation Inc., 2022; Tripathy & Ganapavaram, 2024)

important to build a robust financial foundation. An AAM ecosystem that is *economically self-sustaining* will be able to continuously invest in safety and innovation, weather economic cycles, and steadily expand to serve more communities over time.

RETURN ON INVESTMENT (ROI), COST-BENEFIT, AND ECONOMIC IMPACT OF AAM

From both investor and societal perspectives, AAM must demonstrate a favorable balance of costs and benefits. Significant upfront expenditures are required to develop aircraft, build vertiports, and obtain certifications—raising the question: will the returns justify these investments? For private investors, ROI will be measured in financial returns (profits, equity value) over time. For the public sector, "returns" may manifest as net societal benefits—such as improved mobility, job creation, and economic growth—relative to the public funds invested.

RETURN ON INVESTMENT

Analysts caution that AAM is a long-term play (see Table 10.1). Many AAM manufacturers and operators are still in the pre-revenue stage, and it may take years of operation to recoup their initial investments. An industry study noted that it may take "more than a decade for [the] platform ROI to materialize due to exhaustive R&D and certification stages" (Deloitte, 2023), signaling to early investors that they will

TABLE 10.1
ROI and Market-Size Projections for AAM

Source	Publication Year	Market-Size Estimate (USD)	Target Year	Key Assumption(s)
Morgan Stanley (Base)	2021	$1 trn	2040	Slower early adoption; certification drag
Morgan Stanley (Bull)	2021	$9 trn	2050	Mass urban adoption; autonomy and high utilization
Deloitte (global ROI)	2023	Positive NPV but ROI > 10 years for OEMs	—	Full certification and scale manufacturing
UK Future Flight Challenge	2024	+1.8 % GDP to the United Kingdom (≈ $48 bn)	2030	Nation-wide roll-out; mixed pax and cargo use
Utah AAM Impact Study	2025	$8 bn net economic activity	2045	State-wide vertiport grid; 11,000 jobs
CAAC and MIIT et al. (China)	2024	~¥2 trn (≈ $230 bn) annual output	2030	National low-altitude economy; UAM + UAV logistics

Source: Jonas et al. (2021), Deloitte (2023), UKRI (2024), Utah Department of Transportation (2025), and CAAC and MIIT et al. (2024).

need patience before seeing substantial returns. This trajectory aligns with other disruptive transport technologies (e.g., electric cars), which required sustained investment over 10+ years before reaching profitable scale (Garrow et al., 2021).

The bullish case for AAM is that once aircraft are certified and services begin, the market could grow exponentially—delivering high ROI in the long run. Morgan Stanley (Jonas et al., 2021), for instance, estimated the total market for urban/AAM in the trillions, with their initial projection of a $2.9 trillion global UAM market by 2040, which was later revised to about $1.5 trillion by 2040 (base case), reflecting a more conservative early adoption curve.

Cost-Benefit Analysis

AAM's net benefits need to be evaluated through cost-benefit analysis (CBA), weighing costs (such as vehicle development, infrastructure, noise/airspace management) against anticipated benefits (including time savings for travellers, reduced surface congestion, and improved regional connectivity). A frequently cited benefit is reduced travel time (Rothfeld et al., 2021)—eVTOL air taxis could dramatically cut commute times in gridlocked cities or improve accessibility in regions lacking efficient ground transportation. The monetary value of time saved for businesses and commuters can be significant, bolstering the economic case for AAM.

Improved connectivity can also have secondary benefits, including enhanced productivity, increased tourism, and greater regional economic integration. For example, the UK government's Future Flight Challenge projected that introducing AAM services could increase UK GDP by 1.8% by 2030 (UK Research & Innovation, 2024), due mainly to productivity gains and new business opportunities. Similarly, state-level studies in the United States forecast substantial impacts. A Utah economic analysis found that deploying AAM could create over 11,000 new full-time jobs in that state over two decades and generate more than $8 billion in new business activity (Utah Department of Transportation, 2025). These findings suggest that from a societal perspective, the benefits of AAM could far outweigh its costs—if the technology is implemented effectively. One research review even concluded that AAM's benefits (e.g., time savings, emissions reductions, accidents avoided by shifting trips from roads to air) are *"large and varied, far outweighing its costs"* (Pertz et al., 2023). However, such optimistic conclusions assume high adoption rates and declining unit costs over time. In any case, early cost-benefit studies are helping policymakers identify which AAM use cases (such as airport shuttles or medical deliveries) might deliver positive net value soonest, guiding initial deployments to where they make the most economic sense (McNab, 2024).

Public-Private Partnerships

AAM will require new infrastructure—vertiports, charging stations, and upgraded air traffic management systems. Building this infrastructure is capital-intensive and involves uncertainty due to factors such as demand, regulatory approval, and community acceptance. To tackle this, public-private partnership (PPP) models are being adopted, which share costs and risks between the government and the private sector.

Analyses have emphasized that flexible PPP arrangements can help address uncertainties in AAM by allowing partners to adjust roles as the market evolves (Dulia & Shihab, 2025). We are already seeing examples of PPPs for vertiport development. In Italy, a state-owned airport operator (Atlantia) and an eVTOL company (Volocopter) have partnered to plan an air taxi network in Rome and Venice, sharing investment in vertiport infrastructure (Volocopter, 2022). Japan offers another case: the Osaka prefectural government, in collaboration with private firms, is preparing vertiports and traffic management for demonstrations at Expo 2025 Osaka, aiming to showcase AAM flights during the World Expo (Brinkmann, 2024). In these cases, public authorities provide land, funding, or regulatory support for infrastructure in exchange for the expected public benefits (e.g., bolstering tourism, showcasing innovation).

Case Study: UK Future Flight Challenge

The United Kingdom's comprehensive approach to AAM is worth analyzing. Through the Future Flight Challenge, the government has adopted a holistic strategy, involving funding technology demonstrators, requiring industry co-investment (consortia must include private match funding), and focusing on ecosystem integration (encompassing vehicles, airspace, infrastructure, and public acceptance). One flagship project is the Air Mobility Ecosystem Consortium (AMEC) led by Atkins (UK Research & Innovation, 2024). Backed by a £9.5 million government grant, this consortium is demonstrating end-to-end AAM operations—including a piloted Vertical Aerospace VX4 eVTOL flight from an airport in England to Heathrow, integrated with live air traffic control. The consortium comprises an aircraft OEM (Vertical Aerospace), an airline (Virgin Atlantic, as the operator), an airport operator (Heathrow), a vertiport developer (Skyports), the UK air navigation service provider (NATS), and academic partners. By having all these stakeholders collaborate in a government-funded trial, the United Kingdom ensures that by the time regulations allow commercial AAM flights, the necessary ecosystem (infrastructure, procedures, public confidence) will be ready, significantly lowering the risk and cost that private operators would face if they had to build all this alone. The UK government explicitly ties this initiative to economic goals: "leveling up" regional economies, creating jobs, and achieving net-zero aviation targets. In effect, the United Kingdom is backing the learning phase of AAM to secure a first-mover advantage for its industry and to realize public benefits early. If the demonstrations succeed, British companies could lead the way in AAM, and UK cities could be among the first to benefit from advanced air services, giving the country a competitive edge.

BARRIERS TO FINANCIAL SCALABILITY AND SOLUTIONS

While the potential of AAM is great, so are the financial challenges, as listed in Table 10.2. Scaling up AAM from prototypes and demonstrations to a widely deployed, profitable transportation mode faces several economic barriers. Upheavals, such as the introduction of *trade tariffs*, *geopolitical tensions*, and downturn of the *economic cycle*,

TABLE 10.2
Key Financial Barriers vs. Mitigation Strategies

Barrier	Financial Impact	Lead Stakeholders	Current Mitigation Actions	Example Initiative
Up-front capital (>$500 m per type)	Cash-flow risk; dilution	OEMs, VC/PE	Strategic auto-OEM equity; SPAC proceeds; pre-delivery payments	Toyota-Joby; United $10 m PDAs
Long certification timelines	Delayed revenue	OEMs, investors	Milestone-based finance; dual-use cargo ops	Joby Part 135 ops pre-TC
Vertiport/zoning delays	Stranded assets	Operators, cities	PPPs; vertiport design standard ASTM F3423-23	Skyports-Heathrow demo
Pilot and labor costs	Margin erosion	Operators	Remote ops, autonomy R&D	Wisk Gen 6 autonomous design
Supply-chain scale	High unit cost	OEMs, suppliers	Auto-grade manufacturing JVs; LT supply contracts	Archer-Stellantis factory
Regulatory fragmentation (global vs. China path)	Duplicate compliance cost	OEMs, CAAC, EASA, FAA	Bilateral validation talks; CAAC special condition eVTOL TC	EHang EH216-S's first CAAC type-certificate (2023)

Source: Deloitte (2023), ASTM F3423-23, Archer Aviation (2022), CAAC Press Release (October 2023), and NDRC (2023).

might disrupt both the logistical and financial scalability of the AAM ecosystem. Yet, industry and government players are addressing these challenges and working to overcome them—proving that a financially resilient AAM ecosystem is key to its success.

High Capital Requirements

The most fundamental barrier is the sheer amount of capital required to bring AAM to fruition. Designing, certifying, and manufacturing a new category of aircraft—and concurrently establishing a network of vertiports and operational support—is an enormously expensive endeavor. As noted, experts estimate the development cost per new eVTOL model can range from hundreds of millions up to a few billion dollars (Villamizar, 2025). This level of capital intensity is unusual for startups (especially compared to software or internet ventures). It means AAM firms must continuously raise funds over many years before generating significant revenue. It leaves them vulnerable to shifts in investor sentiment or macroeconomic conditions. For example, Lilium's cash reserves reportedly dwindled from around $205 million in mid-2024 to essentially zero by the end of 2024 as an anticipated funding round failed to close (Villamizar, 2025).

Solution Approaches

AAM companies are addressing this through strategic partnerships and corporate backing—aligning with deep-pocketed partners that can provide capital and long-term support. Joby's partnership with Toyota and Wisk's backing by Boeing are examples that give those startups much longer financial runways than they would have on their own. Another approach is staggered commercialization: focusing on revenue-generating opportunities earlier than full passenger service. As mentioned, some eVTOL developers are pursuing military contracts (which come with government funding) or cargo missions to earn income before passenger certification. Others have sold non-core assets or technology (for instance, selling propulsion battery packs to other industries) to bring in cash. Public sector help can ease capital needs as well—government grants, loans, or purchase commitments provide non-dilutive funding or assured future revenue. And finally, many AAM startups went public via Special Purpose Acquisition Company (SPAC) mergers to tap public equity markets earlier than the traditional Initial Public Offering (IPO) timeline would allow. This generated hundreds of millions of dollars, but at the cost of increased public scrutiny of spending. Those who did so (Joby, Archer, etc.) are now using that cash to secure certification, while those who stayed private are increasingly seeking bridge financing or government aid.

Market Competition

Ultimately, the competitive landscape presents a significant financial challenge. A significant number of companies globally are reportedly developing novel aircraft for urban air mobility (Sengupta et al., 2025). Inevitably, not all will succeed—but if too many reach the market around the same time, there could be an oversupply of services relative to demand in early years, leading to price wars or consolidation that erode profits, as we have already seen some shakeout, as described with Lilium and Volocopter. For investors, the concern is funding a company that another will ultimately outcompete with better technology or deeper pockets.

Solution Approaches

Some degree of *consolidation* is expected and healthy for the industry. Mergers, acquisitions, or partnerships can combine strengths and eliminate redundancies, thereby enhancing overall efficiency. We have seen early moves in this direction: for example, Astro Aerospace (a small player) acquired Horizon Aircraft to bolster its technology (Helicopters Magazine, 2021). The industry appears to be evolving toward a *smaller set of well-capitalized players*—essentially those with the "billion-dollar war chests" to endure—vs. those running on fumes. Those leading players (e.g., Joby, Archer, Beta, and perhaps EHang in China) are likely to capture the first-mover advantage in key markets, making it harder for latecomers to break in unless they have a distinct niche. Many companies are therefore pivoting to *specialize* in niches or B2B roles to avoid head-to-head competition with the leaders. For instance, some might specialize in regional commuter eVTOLs (longer range, 10–15 seaters) instead of urban hops, or focus solely on cargo, or become Tier-1 suppliers of subsystems (such as autonomous flight software) to the leading OEMs, rather than building full vehicles themselves. This type of differentiation enables multiple companies to succeed by serving different segments of what will be a diverse AAM ecosystem.

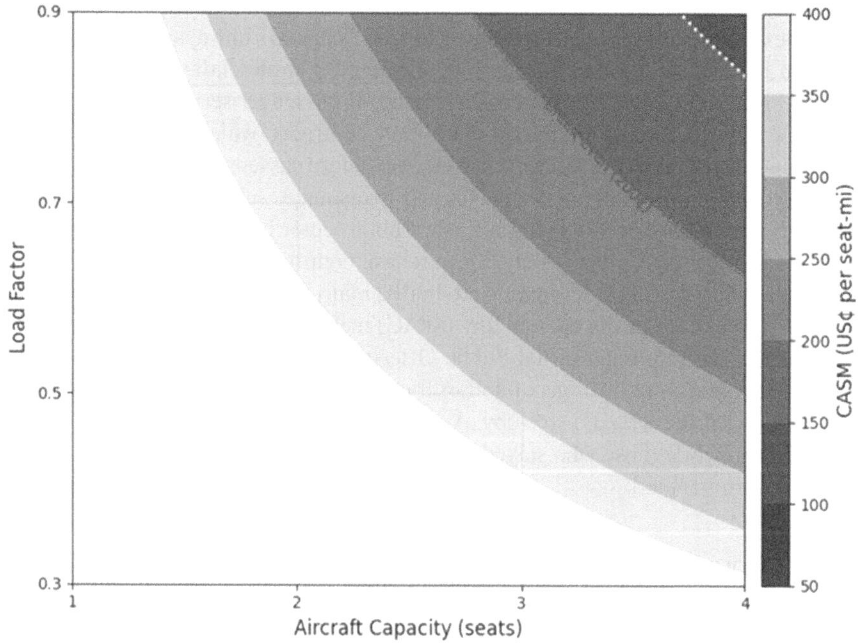

FIGURE 10.3 Cost per Seat-Mile vs Passenger Load Factor. (Credit: Author, Data source: Holden & Goel, 2016; Joby Aviation, Inc., 2021; Meng et al., 2024; Mhlanga, 2019; National Aeronautics and Space Administration [NASA], 2019; Vascik et al., 2024)

Figure 10.3 positions AAM passenger-service unit economics in the same logic used by airlines—linking revenue and cost per available seat-mile (RASM and CASM) to passenger load factor (PLF)—because operating costs are diluted as occupancy and utilization rise. Evidence from airline operations shows that higher load factors are associated with improved cost efficiency (Mhlanga, 2019), and eVTOL economic models similarly treat pooling and high seat utilization as decisive levers, e.g., an average ~67% of revenue-producing seats filled is assumed in a widely cited UAM cost model, broadly consistent with ~70% load factors typical of regional airline operations (Holden & Goel, 2016). Using a ground benchmark, pooled ride-hail fares around US$1.50 per mile have been reported in a major U.S. market (Meng et al., 2024), while an AAM manufacturer's illustrative at-scale economics (CASM ≈ US$0.86 with ~2.3 passengers on average) implies cost per passenger-mile on the order of ~US$1.5 and falling further as load factor increases (Joby Aviation, Inc., 2021). Accordingly, the figure is important because it makes explicit that sustained high occupancy and utilization are not "nice-to-haves" but core prerequisites for eVTOL services to approach affordability and scale demand sustainably (NASA, 2019; Vascik et al., 2024).

In terms of preventing fierce competition in a single city, collaborations might emerge—for example, operators might coordinate through a third-party platform that offers interline booking (so two small operators could serve a city jointly rather than undercut each other). Cities might also initially license a limited number of operators (similar to how some municipalities have handled the rollouts of rideshare and scooter services) to prevent oversaturation.

Price discipline will be crucial; if one deep-pocketed player drastically undercuts fares to capture market share, it could trigger a price war that is beneficial for consumers but detrimental to business. So far, most AAM companies have discussed premium pricing and do not signal an intent to lower prices—they seem to recognize that, in these early days, cooperation in building the market is more valuable than competing on price.

REFERENCES

Archer Aviation. (2022, August 10). *Archer receives $10 million pre-delivery payment from United Airlines for 100 eVTOL aircraft; advances path to commercialization* [Press release].

Brinkmann, P. (2024, November 8). Japan viewed as promising market for advanced air mobility. *Aerospace America*. https://aerospaceamerica.aiaa.org/densely-populated-japanese-cities-viewed-as-perfect-for-electric-air-taxis/

Brunelli, M., Ditta, C. C., & Postorino, M. N. (2023). New infrastructures for Urban Air Mobility systems: A systematic review on vertiport location and capacity. *Journal of Air Transport Management*, 112, 102460. https://doi.org/10.1016/j.jairtraman.2023.102460

Civil Aviation Administration of China (CAAC). (2024, February 29). *China's low-altitude economy exceeds 500b yuan: CAAC*. China Daily.

Deloitte. (2023, December 11). Advanced air mobility: Achieving scale for value realization. *Deloitte Insights*. https://www.deloitte.com/us/en/insights/industry/aerospace-defense/advanced-air-mobility-evtol-aircraft.html?id=us:2el:3dc:aam23ipc:awa:er:010224

Dulia, E. F., & Shihab, S. A. M. (2025). Building public–private partnerships for advanced air mobility infrastructure using game theory. *Journal of Air Transportation*, *33*(3), 229–245. https://doi.org/10.2514/1.D0447

Finnerty, R. (2023, October 26). Beta delivers first electric aircraft to USAF. *FlightGlobal*. https://www.flightglobal.com/fixed-wing/beta-delivers-first-electric-aircraft-to-usaf/155563.article

Garrow, L. A., German, B. J., & Leonard, C. E. (2021). Urban air mobility: A comprehensive review and comparative analysis with autonomous and electric ground transportation for informing future research. *Transportation Research Part C: Emerging Technologies*, *132*, 103377. https://doi.org/10.1016/j.trc.2021.103377

Helicopters Magazine. (2021). Astro Aerospace acquires Horizon Aircraft Inc. *Helicopters Magazine*. https://www.helicoptersmagazine.com/astro-aerospace-acquires-horizon-aircraft-inc/

Holden, J., & Goel, N. (2016). *Fast-forwarding to a future of on-demand urban air transportation* (Uber Elevate white paper). Uber Technologies, Inc.

Joby Aviation, Inc. (2021). *Investor presentation* (Exhibit 99.1 to Form 8-K filed August 11, 2021). U.S. Securities and Exchange Commission.

Jonas, A., Liwag, K. T., Shanker, R., Sharpe, M., Kovanis, B., Silverberg, E., Zaghi, E., & Pasterchick, M. J. (2021, May 6). *Urban Air Mobility: eVTOL/Urban Air Mobility TAM Update: A Slow Take-Off, But Sky's the Limit* (Morgan Stanley Research report). Morgan Stanley & Co. LLC.

McNab, R. M. (2024). Advanced air mobility, economic impacts, and equity considerations. *Journal of Economic Analysis*, *3*(2), 134–146. https://doi.org/10.58567/jea03020009

Meng, S., Brown, A., & Barajas, J. M. (2024). *Complements or competitors? Equity implications of taxis and ride-hail use in Chicago. Journal of Transport Geography, 120*, 103973. doi:10.1016/j.jtrangeo.2024.103973

Mhlanga, O. (2019). Factors impacting airline efficiency in southern Africa: A data envelopment analysis. *GeoJournal, 84*, 759–770. doi:10.1007/s10708-018-9889-9

Ministry of Industry and Information Technology of the People's Republic of China; Ministry of Science and Technology of the People's Republic of China; Ministry of Finance of the People's Republic of China; & Civil Aviation Administration of China. (2024, March 27). *通用航空装备创新应用实施方案 (2024—2030年)* [Implementation plan for innovative application of general aviation equipment (2024–2030)]. (Text reproduced in Hunan Daily/Tencent posting).

National Aeronautics and Space Administration. (2019). *Market study for urban air mobility.* NASA.

National Aviation Authorities Network. (2025, April). Roadmap for advanced air mobility aircraft type certification (Edition April 2025). Federal Aviation Administration. https://www.faa.gov/air-taxis/NAA-Network-Roadmap-Advanced-AirMobility-Aircraft-Type-Certification-Edition-April2025.pdf

National Renewable Energy Laboratory. (2023). Federal Aviation Administration vertiport electrical infrastructure study (NREL/TP-5R00-86245; Revised December 2023). https://docs.nrel.gov/docs/fy24osti/86245.pdf

Pertz, J., Niklaß, M., Swaid, M., Gollnick, V., Kopera, S., Schunck, K., & Baur, S. (2023). Estimating the economic viability of advanced air mobility use cases: Towards the slope of enlightenment. *Drones, 7*(2), 75. https://doi.org/10.3390/drones7020075

Reuters. (2024, November 26). Lilium's fall throws spotlight on air-taxi cash crunch. ETAuto (The Economic Times). https://auto.economictimes.indiatimes.com/news/auto-technology/analysis-liliums-fall-throws-spotlight-on-air-taxi-cash-crunch/115687370

Rothfeld, R., Fu, M., Balać, M., & Antoniou, C. (2021). Potential urban air mobility travel time savings: An exploratory analysis of Munich, Paris, and San Francisco. *Sustainability, 13*(4), 2217. https://doi.org/10.3390/su13042217

S&P Global Market Intelligence. (n.d.). *Airlines industry KPIs for investment professionals.* S&P Global.

Sengupta, R., Bulusu, V., Mballo, C. E., Onat, E. B., & Cao, S. A. (2025). Urban air mobility research challenges and opportunities. *Annual Review of Control, Robotics, and Autonomous Systems, 8*, 407–431. https://doi.org/10.1146/annurev-control-022823-031353

Straubinger, A., Michelmann, J., & Biehle, T. (2021). Business model options for passenger urban air mobility. *CEAS Aeronautical Journal*, 12, 361–380. https://doi.org/10.1007/s13272-021-00514-w

Tripathy, A., & Ganapavaram, A. (2024, November 7). JAL-Sumitomo JV secures right to place order for up to 100 Archer air-taxis. *Reuters.* https://www.reuters.com/business/aerospace-defense/jal-sumitomo-jv-secures-right-place-order-up-100-archer-air-taxis-2024-11-07/

UK Research and Innovation. (2024, July). *Future Flight use cases: 9 ways Future Flight will transform aviation.* Innovate UK & Economic and Social Research Council (ESRC).

U.S. Government Accountability Office. (2022, November). Transforming aviation: Congress should clarify certain tax exemptions for advanced air mobility (GAO-23-105188). https://www.gao.gov/assets/820/813634.pdf

Utah Department of Transportation. (2025). *The economic and community benefits of advanced air mobility* (2-page brief). Utah Department of Transportation, Division of Aeronautics. https://connect.udot.utah.gov/wp-content/uploads/2025/03/NEXA-UTAH-Report-2-pager-Feb-4.-2025-1-1.pdf

Vascik, P. D., Hofer, G., & German, B. (2024). *Ridesharing effects in passenger urban air mobility at scale.* In *Proceedings of the 34th Congress of the International Council of the Aeronautical Sciences (ICAS).*

Villamizar, H. (2025, March 3). Sector analysis: The state of eVTOL innovators. *Airways Magazine.* https://www.airwaysmag.com/new-post/the-state-of-evtol-innovators

Volocopter. (2022). Italy's first vertiport deployed at Fiumicino Airport. *Volocopter Newsroom.* https://www.volocopter.com/en/newsroom/italys-first-vertiport-deployed-at-fiumicino-airport

Part IV

The AAM Outlook

History provides inspiring lessons on how new modes of mobility can rise from fledgling experiments to mainstream reality. A century ago, the idea of routine air travel seemed like a fantasy. On New Year's Day 1914, the world's first regularly scheduled heavier-than-air airline launched in Florida, using Benoist airboats (early seaplanes) to carry passengers across Tampa Bay (Reilly, 1996; IATA, n.d.). The route was about 20 miles (≈21 miles in Reilly's account) and, while daily operations were short-lived, the service continued in some form until 5 May 1914, demonstrating the basic feasibility of scheduled commercial flight (Reilly, 1996; IATA, n.d.). Many observers were skeptical about whether aviation could ever become safe or affordable at scale. Yet aircraft builder Thomas W. Benoist prophetically declared his vision: "*Someday people will be crossing oceans on airliners like they do on steamships today*" (IATA, n.d.). At the time, this claim sounded outrageous. However, within about 25 years, transoceanic airliners were indeed a reality, and by the latter half of the 20th century, flying had become one of the safest and most commonplace ways to travel long distances.

The evolution of aviation from 1914 to the Jet Age highlights a pattern we see repeatedly: early innovations start minor and face setbacks, but they lay the groundwork for exponential growth and widespread adoption. The same pattern can be observed in more recent parallels. The electric car revolution: in its early years, EVs were expensive, had limited range, and lacked charging infrastructure. Adoption was slow, and many doubted they could compete with gasoline cars. Yet relentless improvements in battery technology—especially sustained cost declines—and supportive public policy, including purchase incentives and the build-out of reliable charging networks, have materially accelerated EV adoption (Hardman et al., 2017; Hardman et al., 2018; Rezvani et al., 2015; Sierzchula

DOI: 10.1201/9781003546306-14

FIGURE IV.1 World's first scheduled airline service, 1914. (Credit: Shankar S., *Source*: Flickr.)

et al., 2014). In some countries, electric cars account for well over half of new-car sales (e.g., 88.9% of new passenger cars in Norway in 2024), and globally EV sales share has continued to rise year-on-year, reaching about 18% of new car sales in 2023 and exceeding 20% in 2024 (BloombergNEF, 2024; IEA, 2024, 2025; GRIDSERVE, 2025; Norwegian EV Association, 2024).Public perception shifted from viewing EVs as a niche eco-gadget to seeing them as the inevitable future of automobiles. In much the same way, AAM today is in its infancy—aircraft are still being refined, regulations are nascent, and initial services will likely be costly. But each small success (a safe demo flight, a useful delivery mission, a satisfied passenger) builds momentum and public trust. Gradually, the conversation shifts from "*Is this crazy?*" to "*How did we ever live without this?*"

Another illuminating parallel is the digital transformation accelerated by the COVID-19 pandemic. Before 2020, technologies like video conferencing, telemedicine, and remote work were gradually growing but still secondary to traditional in-person norms. The pandemic upended that overnight. In a matter of months, hundreds of millions of people embraced digital solutions out of (Agostino et al., 2021; Soto-Acosta, 2020; Wade & Shan, 2020). Video collaboration illustrates the scale of this shift: Zoom reported that daily meeting participants rose from a peak of 10 million in December 2019 to 300 million by April 2020 (Lawler, 2020; Yuan, 2021). This 30-fold surge was a societal leap that no one anticipated happening so fast. Yet when circumstances demanded it, adoption skyrocketed and habits changed, likely permanently in many cases. The lesson here is powerful: transformative change can happen rapidly once a tipping point is reached. Advanced air mobility (AAM) could experience a similar inflection if external factors (such as worsening urban congestion or urgent climate action) press the need for new transportation options. When AAM services prove their value—say, by dramatically cutting a commuter's travel time or providing critical deliveries in a disaster—public demand could accelerate beyond what current linear forecasts predict. Society has shown an ability to adapt quickly when benefits are clear and the status quo is untenable.

Bringing together these threads, we arrive at a vision of AAM's future that is inspiring yet grounded in responsibility. By 2030, we expect AAM to no longer be an experimental sideshow, but an industry offering extended services in select regions—a new branch of aviation growing in parallel with traditional airlines and ground transport.

However, the true measure of success will be how safely and sustainably this integration happens. The AAM industry has the benefit of learning from aviation's past mistakes and successes. Building a framework for resilient and safe operations from the outset is not just an academic ideal, but a practical necessity. This means upholding a culture where safety is not compromised for the sake of speed of rollout, where data and risk awareness guide decisions as much as ambition does. *Manufacturers* must continue to test rigorously and be transparent about their progress. *Regulators* should remain agile, updating rules as technology advances, but without compromising on certification standards. *Community* engagement is not a box to check; it should be an ongoing dialogue that fosters a sense of ownership and trust in this new mode of mobility for people and goods. *Environmental* impacts need to be monitored and mitigated, ensuring AAM truly delivers on its promise of cleaner transport and does not introduce new problems, such as noise pollution, unchecked.

If these pieces come together, the pathway for AAM is extremely promising. We can imagine moving into the future with AAM networks expanding between and within cities, complementing high-speed trains and electric vehicles to form a seamless intermodal system. AAM will not replace other forms of transportation wholesale—just as airplanes did not replace cars for short trips—but it will fill crucial gaps and create new possibilities. *Rural areas* may gain air connectivity without requiring expensive new roads. *Emergency responders* might rely on drone fleets for rapid aid. *Travelers* may enjoy a breathtaking aerial ride to the airport, turning commutes into minutes of joy rather than stress. And *future students of aviation* will read about our current era as the time when flying truly went electric, became democratized, and decentralized.

AAM's future is bright. It stands on the shoulders of giant achievements in aerospace and leaps into new technological territory. By being *cautiously optimistic*—dreaming big while planning wisely—AAM is mapping a course to a future where the skies could be accessible for all. The question is not *if* AAM will transform how we move, but *when* and *how*. With resilient safety practices, sustainable planning, and a collaborative approach, AAM can mature into a thriving, responsible industry. It is an evolution unfolding before our eyes, and its next chapters will undoubtedly be as challenging and exhilarating as the aviation milestones that came before.

11 Trends and Opportunities

INTRODUCTION

Advanced air mobility (AAM) encompasses a range of revolutionary aerial transport services—from electric passenger air taxis and autonomous cargo drones to public service and emergency response aircraft. By the year 2030, AAM operations are envisioned to be integrated into multimodal transportation networks worldwide, augmenting existing ground and air transport systems (Goyal et al., 2021; Guo et al., 2024; Tuchen et al., 2022). Industry forecasts suggest that thousands of eVTOL aircraft could be flying above cities in the 2030s, offering rapid urban transit, regional connectivity, and on-demand logistics (Heineke et al., 2021). This growth is driven by technological advances in electrification, automation, and digital infrastructure, coupled with a strong push for sustainable and efficient transportation. The goal of this part is to envision how AAM can develop effectively, efficiently, sustainably, resiliently, and, ultimately, more safely than the current aviation industry—pushing the boundaries to what is possible in air transport.

SCALING SUSTAINABLY

The AAM industry is projected to grow significantly in the 2030s, but scaling this ecosystem sustainably requires a careful balance of innovation with safety and community concerns. One central challenge is ensuring *sustainable development*—not only in the environmental sense but also in economic and social terms, as presented in Part III. Parallel to economic and environmental sustainability is the *safety sustainability* of AAM operations. Aviation's safety culture—honed over a century—must carry into AAM from day one. This means robust design standards, pilot (or operator) training, airspace integration, and Safety Management Systems (SMS) adapted to the nuances of AAM. Regulators, such as the FAA and EASA, have signaled that existing rules will evolve to accommodate novel vehicles and higher traffic densities without compromising safety. For instance, the FAA has been developing concepts like Unmanned Traffic Management (UTM) and Extensible Traffic Management (xTM) to handle autonomous and high-volume operations at low altitudes. New categories of flight operations are emerging with the *Autonomous Flight Rules (AFR)* as a counterpart to Visual and Instrument Flight Rules (Wing & Cotton, 2011). Under AFR, autonomous or highly automated flights would follow a defined set of rules and procedures, allowing them to safely blend into airspace alongside VFR and IFR traffic. By establishing such rules, authorities aim to maintain the same level of safety as today's system while enabling the benefits of automation.

DOI: 10.1201/9781003546306-15

Finally, public acceptance is a critical pillar of *sustainable* AAM growth. Surveys indicate that many consumers are intrigued by air taxis but will demand evidence of safety and noise mitigation. Early AAM services will likely start with human pilots on board (or monitoring remotely) to build trust. Over time, as automation proves itself, operations may shift to *pilot-optional* or fully autonomous modes. Throughout this evolution, maintaining a *resilient safety focus*—one that proactively identifies and mitigates risks—will be paramount.

AAM Adoption Scenarios

Several scenarios could be suggested based on the current available information for the pathway of adoption of AAM at a global scale. As depicted in Figure 11.1, three adoption scenarios—Low, Base, and High—each broken down into Cargo, Passenger, and Public Service fleets using a logistic growth model. The Low scenario caps at ~50,000 total aircraft by 2040, reflecting conservative adoption assumptions, while in the Base scenario the number of in-service aircraft reaches ~200,000 by 2040, and last the High scenario scales to ~500,000 by 2040, representing an aggressive growth path noted in industry (Abermann, 2025; Bank of America Global Research, 2025; Chan & Lui, 2025; China Daily, 2024; Gao & Huang, 2021; Ibeawuchi, 2025; Morgan Stanley Research, 2019; World Economic Forum, 2024).

EMBRACING AUTONOMY

One of the defining features of AAM is the high degree of autonomy envisioned in vehicle operations and traffic management. Autonomy is seen as the key to making AAM scalable, safe, and efficient, effectively acting as a "smart copilot" that is always alert (Skygrid, 2025). Without autonomy, AAM's scale would be infeasible, as humans in the loop for every decision would quickly become a bottleneck. By delegating routine piloting tasks and strategic traffic coordination to algorithms, AAM aircraft can maintain safe distances, optimize routes, and respond to hazards in real time without constant human intervention. This autonomy, however, introduces new challenges for safety assurance.

As the industry asks how autonomy will be able to alter long-held aviation practices, several forward-looking concepts have emerged:

Autonomy Factors Analysis and Classification System (AFACS)

In traditional aviation, investigators often use the Human Factors Analysis and Classification System (HFACS) to categorize the human errors contributing to accidents (Wiegmann & Shappell, 2003). However, AAM vehicles (especially drones and autonomous eVTOLs) will shift some of this risk to autonomous systems. There is a growing recognition that accident analysis frameworks must evolve to capture not only human errors but also failures in automation, software, and sensors. Researchers analyzing unmanned aircraft mishaps have noted that HFACS, as originally formulated, *"only considers human factors as causes and does not include automation, manufacturing malfunctions, and system or communication failures"*

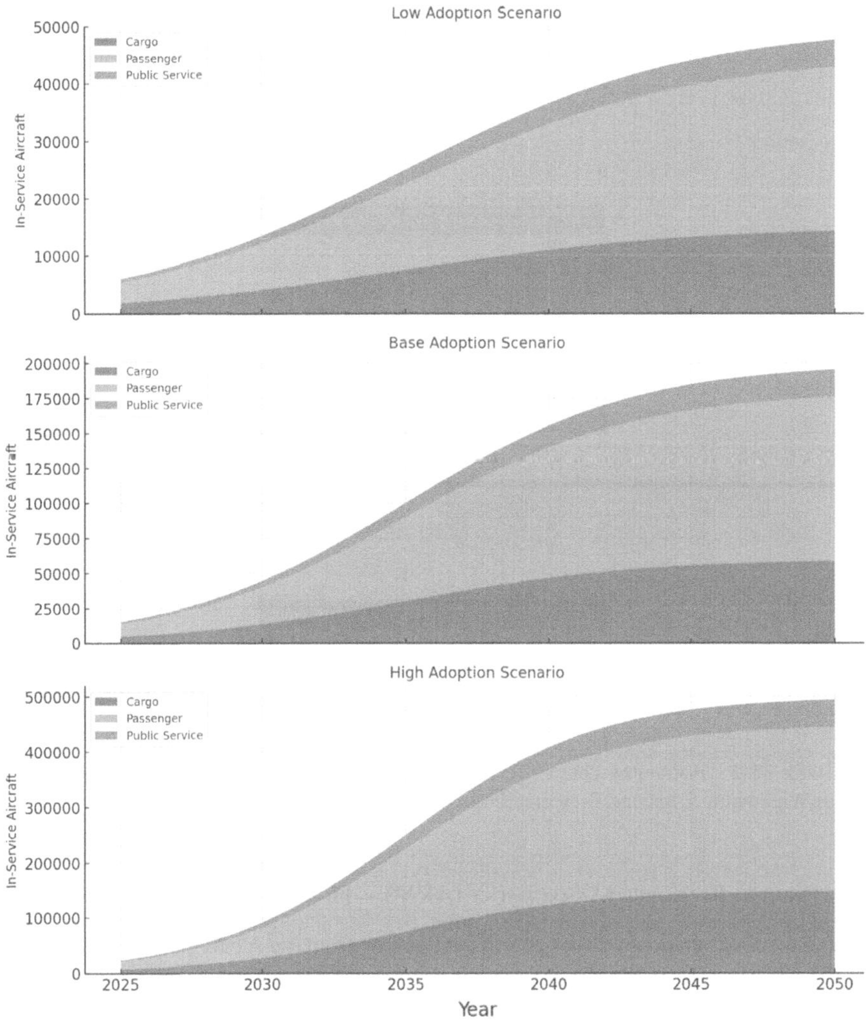

FIGURE 11.1 Global AAM S-curve (2025–2050).

(El Safany & Bromfield, 2025). Given that many UAV accidents can be attributed to issues such as automation logic errors or loss of communication links, a proposed *Autonomy Factors Analysis and Classification System* (*AFACS*) could help fill this gap, as illustrated in Figure 11.2. *AFACS* would systematically classify how and why an autonomous system contributed to an incident—for example, distinguishing between a software decision error, a sensor failure, or an unforeseen algorithmic interaction. By learning from autonomy-related mishaps (even minor incidents), the AAM industry can refine vehicle algorithms and interfaces to prevent recurrences. In essence, just as HFACS improved understanding of human error, *AFACS* would improve understanding of *machine or autonomy error*, enabling safer designs and better oversight of autonomous systems (see Table 11.1).

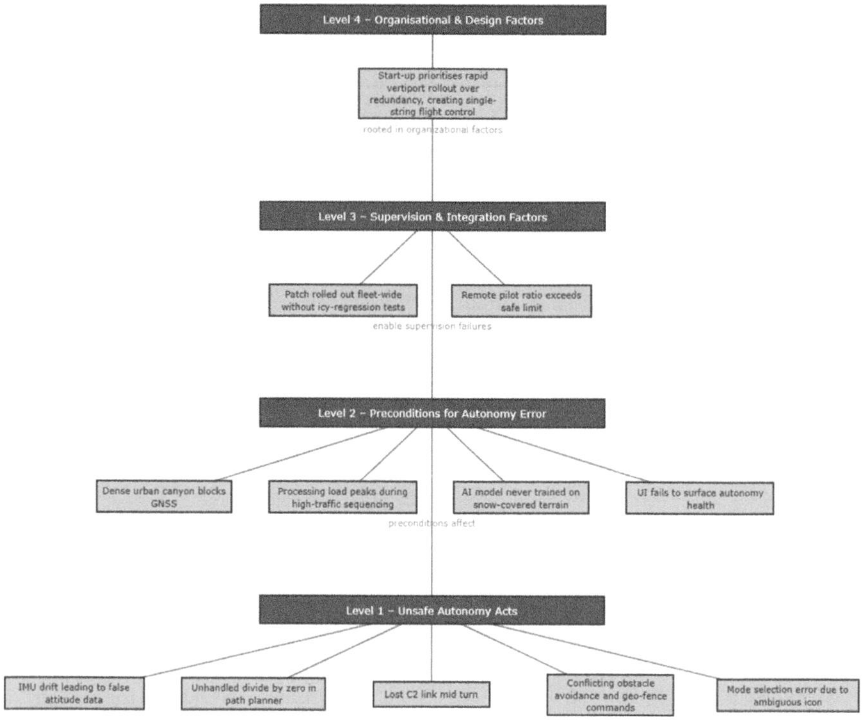

FIGURE 11.2 Autonomy Factors Analysis and Classification System (AFACS). (Adapted from Wiegmann & Shappell's work, 2003).

Augmented Resource Management (ARM)—Human-Autonomy Teaming

The rise of automation does not eliminate humans from the AAM ecosystem; rather, it reshapes their role. Pilots, ground operators, maintenance crews, and air traffic managers will increasingly work *with* autonomous systems in a collaborative way. This has led to the concept of extending Crew Resource Management (CRM) into *Augmented Resource Management (ARM)* for human-machine teams. Traditional CRM training focuses on communication, teamwork, and decision-making among human crew members to minimize errors and enhance overall performance. In AAM, the "crew" may include onboard AI copilots, remote supervisors, and automated decision-support tools. The question is how to manage all available resources—both human and machine—to assure a safe operation. Studies of large unmanned aircraft systems (UAS) have shown that CRM principles (like teamwork, communication, and workload management) remain essential for safety even when the aircraft is remotely piloted or autonomous (Neff, 2019). Building on this, *ARM* would allow training of human operators to effectively oversee and trust autonomous functions, while also training the designers of autonomous agents to ensure their systems are transparent and supportive team players.

Human-autonomy teaming research indicates that clear allocation of authority and responsibility between humans and AI is vital—the human should know *when*

TABLE 11.1
AFACS Framework

AFACS Level	Category	Typical Failure Modes	Illustrative Incident Example	Representative Mitigations
Level 1—Unsafe Autonomy Acts	• Sensor/Hardware Failure • Software Logic/Algorithm Error • Communication/Link Loss • Autonomy-Decision Anomaly • Human-Automation Interface Slip	IMU drift leading to false attitude data. Unhandled divide-by-zero in path-planner, Lost C2 link mid-turn • Conflicting obstacle-avoidance and geo-fence commands • Mode-selection error due to ambiguous icon	eVTOL abruptly pitches nose-down after inertial-sensor freeze; autopilot misclassifies attitude, triggering aggressive correction	Real-time sensor cross-checks; fail-operational flight-control computers; deterministic software verification; link redundancy; interface re-design with distinct mode cues
Level 2—Preconditions for Autonomy Errors	• Degraded Sensing Environment • Computing Resource Exhaustion • Latency/Bandwidth Constraints • Model Training Gaps • Operator Situational-Awareness Erosion	Dense urban canyon blocks GNSS; processing load peaks during high-traffic sequencing; the AI model has never been trained on snow-covered terrain; the UI fails to surface autonomy health.	Multisensor fusion fallback; onboard resource-monitor watchdogs; adaptive QoS for datalinks; continuous-learning datasets; SA dashboards with automation-status cues	
Level 3—Supervision and Integration Factors	• Inadequate Autonomous-System Testing • Insufficient Human Oversight Protocols • Poor Configuration/Change Control • Weak Safety-Assurance Processes	Software patch rolled out fleet-wide without regression test in icy-runway scenarios; remote pilot ratio exceeds safe limit.	Rigorous scenario-based test suites; oversight workload thresholds; version-control gates; real-time safety-performance monitoring	
Level 4—Organizational and Design Factors	• Policy Gaps for Autonomy Certification • Resource Constraints on Cyber-Resilience • Misaligned Safety Culture (Human vs. Machine) • Inadequate Training Doctrine for AI • Economic Pressure Over Safety	Start-up prioritizes rapid vertiport rollout over redundancy investment, leading to a single-string flight control architecture.	Regulatory standards for AFR operations; budget earmarks for resilience; unified "autonomous safety" culture programs; continuous upskilling for AI stewardship	

to intervene, and the AI should know *when* to defer to human judgment. One can imagine future AAM pilot training that includes managing fleets of semi-autonomous drones or working alongside an AI that handles routine navigation, allowing the pilot to focus on strategic decisions.

Conversely, AI systems are being designed to understand human intentions and to provide timely alerts or suggestions (much like a human co-pilot would). The end result of ARM would be a seamless human-machine team that capitalizes on the strengths of each: the consistency, multitasking capabilities, and tirelessness of automation, combined with the creativity and moral judgment of humans.

Augmented Resource Management (ARM) Model for Human-Autonomy Teaming in AAM

ARM extends classic CRM into a full human-autonomy-team construct. The proposed model—ARM-T (Transparency) A (Authority) S (Situational Awareness) T (Trust) W (Workload), or *T-A-S-T-W* framework (see Table 11.2)—is scientifically grounded in peer-reviewed human-autonomy-teaming research (e.g., Parasuraman et al., 2000; Chen & Barnes, 2014; Endsley, 2017; Lee & See, 2004) and recent NASA AAM studies (Holbrook et al., 2020). Each pillar includes measurable constructs, training objectives, and design levers that together yield resilient, safe AAM operations.

How ARM-TASTW Operates

In the design phase, engineers embed transparency hooks—such as explainable-AI modules—alongside authority-ladder logic and workload-sensing capabilities within avionics and ground control stations. During the training phase, operators complete scenario-based drills that emphasize authority negotiation and trust calibration, while AI agents are conditioned for cooperative behaviors (Holbrook et al., 2020). In the operational phase, real-time monitors track workload, trust, and situational awareness indicators so adaptive logic can adjust the level of autonomy or proactively surface explanations. Finally, in the evaluation and SMS loop, post-mission data feed autonomy-risk-management metrics into the organization's SMS to drive continuous improvement (Neff, 2019).

SCIENTIFIC VALIDATION PATH

- Simulation trials using NASA's HART scenarios to compare T-A-S-T-W-trained crews versus baseline CRM crews for multi-eVTOL supervision (expected ≥ 20 % reduction in SA-loss events).
- Human-in-the-loop experiments with variable autonomy levels to confirm authority-handoff safety (cf. Kaber & Endsley, 2004).
- Field pilots in Urban Air Mobility testbeds (e.g., FAA AAM corridors) collecting workload, trust, and performance data; statistical analysis following Chen and Barnes (2014) methodology.

By aligning each pillar with established human-factors constructs and measurable safety outcomes, the ARM-TASTW framework offers a scientifically

TABLE 11.2
ARM-TASTW Framework

ARM Pillar	Scientific Basis	Key Design/Training Objectives	Illustrative Mechanisms for AAM	Representative Metrics
Transparency (system legibility)	• XAI and explainability research (Gunning & Aha, 2019) • HAT design-pattern work (Shively et al., 2016)	Provide real-time *why/what-next* displays. Support, and *explain-back* queries from the crew	• Interactive "explain" button overlays on eVTOL flight path • Voice-prompted rationale for conflict-avoid resolutions	• Time-to-comprehend automation intent • Surveyed transparency usability (Likert)
Adaptive Authority (dynamic control hand-offs)	• Levels of Autonomy model (Parasuraman et al., 2000) • Dynamic delegation studies (Kaber & Endsley, 2004)	Pre-define trigger criteria for role shifts. Train crews on *when/how* to take back control.	• "Authority ladder" in ground station UI, highlighting current autonomy level • Rule-based hand-back if sensor confidence < 0.7	• Handover success rate • Intervention timeliness • Authority-mode error count
Shared Situational Awareness (SA)	• SA theory (Endsley, 1995) • Multi-UAV SA findings (Cummings and Mitchell, 2007)	Maintain a common mental model across humans and AI. Fuse multi-sources of data into succinct displays.	• 360° synthetic vision shared by pilot and AI copilots • Cloud dashboard for remote supervisors	• SAGAT/3D-SART SA scores • Loss-of-SA incident rate
Trust Calibration	• Trust in Automation framework (Lee & See, 2004) • NASA trust research for AAM (Chancey & Politowicz, 2020)	Avoid over- and under-trust (misuse/disuse). Provide continual trust cues and feedback loops.	• Confidence bar linked to autonomy self-health diagnostics • Training scenarios with intentional calm and fault-injection	• Trust questionnaire (Muir scale) • Misuse/disuse frequency
Workload and Team Performance	• Multiple UAV workload studies (Chen & Barnes, 2014) • NASA Single-Pilot Ops workload limits (Bailey et al., 2017)	Balance cognitive load < 75 % of capacity. Use adaptive automation to shed load peaks.	• Automated task shedding: AI handles communications when pilot workload > 0.7 NASA-TLX • Predictive alerts for "crew SA drop"	• NASA-TLX, ISA scores • Oper. error vs. workload curve

defensible roadmap for integrating human and autonomous resources in future AAM operations.

Processor Fatigue Management

Human pilots have well-known limitations—they get fatigued, distracted, or stressed—which is why duty hour limits and fatigue management programs are key safety tools. Autonomous systems, on the other hand, do not get tired or distracted in the human sense, and they execute tasks with relentless consistency. This is a major advantage: automation can eliminate errors caused by human fatigue or lapses in attention. However, automated systems have their own failure modes that must be managed. Just as a human pilot's performance degrades with exhaustion, an autonomous system's performance might degrade due to sensor wear, software bugs, or data anomalies. For instance, continuous operation can lead to issues such as overheating hardware, memory leaks in software (sometimes referred to as *software aging*), or the accumulation of small sensor calibration errors.

In AAM, a new concept of *processor fatigue management* may emerge, referring to the monitoring of the health and uptime of critical onboard computers and algorithms. This might involve scheduling regular resets or self-diagnostics for software, much as humans take rest breaks. It could also mean using redundant systems that periodically cross-check each other, ensuring that if one autopilot computer has slowed or "hung" in a faulty logic loop, another can take over—analogous to a rested co-pilot taking the controls.

Predictive maintenance will also play a role: AAM operators will leverage data to predict when components (from batteries to rotors to sensors) are likely to fail or degrade, and replace them before they cause an incident. In essence, while we won't worry about a computer being sleepy, we will worry about it running beyond its design limits or encountering scenarios it wasn't trained for. Therefore, future SMS and operational protocols may treat the autonomous system itself as an entity that requires care and monitoring. Some measures could include real-time algorithm performance monitoring (to detect anomalies in decisions), periodic software updates and validation tests, and establishing *go/no-go* criteria for the health of critical automation (similar to how a pilot must be medically fit for duty). Embracing this proactive stance—ensuring the *machine* is in a fit state to fly—will be another facet of maintaining high safety in AAM.

AAM AS THE INCUBATOR OF AVIATION 5.0

The confluence of technologies and concepts in AAM is setting the stage for *Aviation 5.0*—a new era of aviation characterized by high levels of autonomy, digital connectivity, sustainable propulsion, and novel operational models. AAM can be viewed as an incubator for these innovations on a smaller scale before they scale up to mainstream aviation. Historically, significant technological shifts in aviation (from propeller to jet propulsion, or the introduction of fly-by-wire) initially occurred in limited contexts and then expanded. Similarly, the advances made in the AAM domain in the 2020s and 2030s could transform how all aircraft are designed and operated in the 2040s and beyond.

Several examples illustrate this symbiosis between AAM and the broader aviation industry as follows.

Autonomy and Pilotless Operations

AAM vehicles are spearheading efforts in pilotless flight. Some eVTOL developers (e.g., Wisk Aero) are pursuing fully autonomous air taxi services from the outset, with no onboard pilot, relying solely on remote human supervision. This aggressive approach, while facing certification challenges, is forcing regulators to devise approval pathways for autonomous passenger flight. The procedures, fail-safes, and regulatory frameworks created for these smaller vehicles will provide a blueprint for larger aircraft. Today, the idea of a *pilotless airliner* still faces technical and public acceptance hurdles. However, if by the 2030s thousands of AAM flights have been conducted safely by automated systems, public confidence in autonomy will grow. We may first see cargo aircraft adopting pilotless operation (since they pose less public risk), followed by regional passenger planes, and eventually long-haul airliners with no human in the cockpit. The timeline is uncertain—for instance, Europe's EASA has ruled out single-pilot airline operations until at least 2030, citing safety concerns—but the direction is set (Freed & Lampert, 2023). By demonstrating autonomy at scale and under real-world stresses, AAM will accelerate the development of larger autonomous aircraft, with the concept of a digital co-pilot that can assume the duties of a second human pilot essentially being prototyped in AAM vehicles today.

Propulsion and Sustainable Technologies

AAM's emphasis on electric or hybrid-electric propulsion is driving rapid development of battery technology, electric motors, and even hydrogen fuel cells for aviation. These technologies, proven on smaller aircraft, can be scaled up to larger planes. For example, improved battery energy densities achieved for eVTOLs will benefit electric general aviation aircraft and could eventually make hybrid-electric regional aircraft viable. The supply chain and infrastructure for electric aviation (charging stations, vertiports with solar panels, battery recycling programs) established for AAM will lay the groundwork for broader adoption.

Furthermore, noise-reduction technologies developed to make AAM vehicles neighborhood-friendly (such as ultra-quiet rotors or advanced acoustic mitigation) might be applied to conventional aircraft to reduce community noise around airports. In this way, AAM accelerates the aviation industry's push toward its sustainability and climate goals by serving as a real-world laboratory for green tech.

Business Models and Multimodal Integration

AAM is introducing new business and service models, such as aerial ride-sharing, drone delivery-as-a-service, emergency response, and vertiport hubs linking with public transit. As these models mature, they could influence conventional aviation. For instance, vertiports in city centers might become extensions of airports, streamlining the first/last-mile connectivity for airline passengers (shaving time off door-to-door travel). Airlines could partner with AAM operators to offer fly-drive packages, or even operate eVTOL fleets themselves for regional hops. Conversely, airport design might change to accommodate AAM traffic (adding vertiport pads on terminal roofs). Through such integrations, the boundary between traditional aviation and AAM blurs, creating a more holistic air transport system. This aligns with the vision of future "smart cities" where various transportation modes—autonomous cars, high-speed trains, and AAM—all coordinate to move people efficiently.

In all these ways, AAM serves as both *student and teacher* to the legacy aviation industry: it learns from the past (applying aviation's hard-won safety practices), and it teaches new approaches for the future (demonstrating what autonomy, electrification, and connectivity can do). This two-way exchange is the core of *Aviation 5.0*. The main opportunity of AAM lies not just in providing new services, but in catalyzing a transformation of the broader aviation sector to be more *effective, efficient, sustainable, resilient, and safe*. If AAM can prove that point-to-point air mobility can be done *autonomously*, with *minimal emissions*, *zero accidents*, and *acceptable cost* (which could vary in a similar approach to the commercial aviation industry: from Low-Cost Carriers business models, to attract the broad market, to Full-Service Carrier models for more niche customer markets), it will open new horizons. We may envision a future where short-haul flights are predominantly autonomous, where pilots manage multiple vehicles from a control center, and where the global airspace operates with an intelligent automated backbone that was born out of today's drone traffic management systems.

REFERENCES

Abermann, N. (2025, June 7). Tesla's next frontier? Analyst sees it ruling the 'low-altitude economy'. *Wall Street Pit*. https://www.wallstreetpit.com/

Agostino, D., Arnaboldi, M., & Diaz Lema, M. (2021). New development: COVID-19 as an accelerator of digital transformation in public service delivery. *Public Money & Management, 41*(1), 69–72. https://doi.org/10.1080/09540962.2020.1764206

Bailey, R. E., Kramer, L. J., Kennedy, K. D., Stephens, C. L., & Etherington, T. J. (2017). *An assessment of reduced crew and single pilot operations in commercial transport aircraft operations*. In IEEE/AIAA 36th Digital Avionics Systems Conference (DASC).

Bank of America Global Research. (2025, June 3). *The "Low-Altitude" Economy is taking off*. Bank of America.

BloombergNEF. (2024, December 10). *Lithium-ion battery pack prices see largest drop since 2017, falling to $115 per kilowatt-hour*. BloombergNEF.

Chan, A., & Lui, S. (2025, May). *Securing take-off for the low-altitude economy: Advanced Air Mobility in Hong Kong and the Greater Bay Area*. AECOM Without Limits.

Chancey, E. T., & Politowicz, M. (2020). *Designing and training for appropriate trust in increasingly autonomous advanced air mobility operations: A mental model approach (Version 1)* (NASA/TM-20205003378). NASA Langley Research Center.

Chen, J. Y. C., & Barnes, M. J. (2014). Human–agent teaming for multirobot control: A review of human factors issues. *IEEE Transactions on Human-Machine Systems, 44*(1), 13–29. https://doi.org/10.1109/THMS.2013.2293535

China Daily. (2024, March 5). *Breaking barriers: Accelerating the future of Advanced Air Mobility*. ChinaDaily.com.

Cummings, M. L., & Mitchell, P. M. (2007). Scheduling the human supervisor: How to schedule humans to monitor multiple unmanned vehicles. *Aerospace Science and Technology*. https://doi.org/10.1016/j.ast.2006.10.007

El Safany, R., & Bromfield, M. A. (2025). A human factors accident analysis framework for UAV loss of control in flight. *The Aeronautical Journal, 129*(1337), 1723–1749. https://doi.org/10.1017/aer.2025.24

Endsley, M. R. (1995). Toward a theory of situation awareness in dynamic systems. *Human Factors, 37*(1), 32–64. https://doi.org/10.1518/001872095779049543

Endsley, M. R. (2016). From here to autonomy: Lessons learned from human–automation research. *Human Factors: The Journal of the Human Factors and Ergonomics Society, 59*(1), 5–27. https://doi.org/10.1177/0018720816681350

Freed, J., & Lampert, A. (2023, February 7). European regulator rules out single-pilot flying by 2030. *Reuters.* https://www.reuters.com/business/aerospace-defense/european-regulator-rules-out-single-pilot-flying-by-2030-2023-02-06/

Gao, Y., & Huang, Z. (2021). Advanced air mobility: Demand analysis and market potential of the personal air vehicle service. *Sustainability, 13*(13), 7421. https://doi.org/10.3390/su13137421

Goyal, R., Reiche, C., Fernando, C., & Cohen, A. (2021). Advanced air mobility: Demand analysis and market potential of the airport shuttle and air taxi markets. *Sustainability, 13*(13), 7421. https://doi.org/10.3390/su13137421

GRIDSERVE. (2025, February 25). *Why Norway leads the world in EV adoption.*

Gunning, D., & Aha, D. W. (2019). DARPA's explainable artificial intelligence (XAI) program. *AI Magazine, 40*(2), 44–58. https://doi.org/10.1609/aimag.v40i2.2850

Guo, J., Chen, L., Li, L., Na, X., Vlačić, L., & Wang, F. Y. (2024). Advanced air mobility: An innovation for future diversified transportation and society. *IEEE Transactions on Intelligent Vehicles, 9*, 3106–3110. https://doi.org/10.1109/TIV.2024.3377464

Hardman, S., Chandan, A., Tal, G., & Turrentine, T. (2017). The effectiveness of financial purchase incentives for battery electric vehicles – A review of the evidence. *Renewable and Sustainable Energy Reviews, 80*, 1100–1111. https://doi.org/10.1016/j.rser.2017.05.255

Hardman, S., Jenn, A., Tal, G., Axsen, J., Beard, G., Daina, N., ... Witkamp, B. (2018). A review of consumer preferences of and interactions with electric vehicle charging infrastructure. *Transportation Research Part D: Transport and Environment, 62*, 508–523. https://doi.org/10.1016/j.trd.2018.04.002

Heineke, K., Kloss, B., & Riedel, R. (2021, November 19). *Advanced air mobility in 2030.* McKinsey & Company. https://www.mckinsey.com/industries/aerospace-and-defense/our-insights/advanced-air-mobility-in-2030

Holbrook, J. B., Prinzel, L. J., & Chancey, E. T., Shively, R. J., Feary, M., Dao, Q. V., Ballin, M. G., & Teubert, C. (2020). *Enabling urban air mobility: Human–autonomy teaming research challenges and recommendations* (NASA TM–20205002387). National Aeronautics and Space Administration. https://ntrs.nasa.gov/citations/20205002387

Ibeawuchi, B. (2025, June 6). *Essential reading for low-altitude economy and Advanced Air Mobility enthusiasts.* BusinessAviation.aero. https://businessaviation.aero/evtol-news-and-electric-aircraft-news/advanced-air-mobility/essential-reading-for-low-altitude-economy-and-advanced-air-mobility-enthusiasts/

International Air Transport Association. (n.d.). *The story of the world's first airline.* IATA.

International Energy Agency. (2024). *Global EV Outlook 2024: Trends in electric cars.* IEA.

International Energy Agency. (2025). *Global EV Outlook 2025* (Report PDF). IEA.

Kaber, D. B., & Endsley, M. R. (2004). The effects of level of automation and adaptive automation on human performance, situation awareness and workload in a dynamic control task. *Theoretical Issues in Ergonomics Science, 5*(2), 113–153. https://doi.org/10.1080/1463922021000054335

Lawler, R. (2020, June 2). Zoom usage peaked at 300 million daily participants in April. *Engadget.*

Lee, J. D., & See, K. A. (2004). Trust in automation: Designing for appropriate reliance. *Human Factors, 46*(1), 50–80. https://doi.org/10.1518/hfes.46.1.50_30392

Morgan Stanley Research. (2019). *Urban air mobility eVTOL/Urban air mobility TAM update: A slow take-off, but sky's the limit.* Morgan Stanley.

Neff, P. S. (2019). Crew resource management for large unmanned aircraft systems operations. *International Journal of Aviation, Aeronautics, and Aerospace*, *6*(3), 1–17. https://commons.erau.edu/ijaaa/vol6/iss3/1/

Norwegian EV Association. (2024). *Norwegian EV market* (market share statistics; updated Dec 31, 2024).

Parasuraman, R., Sheridan, T. B., & Wickens, C. D. (2000). A model for types and levels of human interaction with automation. *IEEE Transactions on Systems, Man and Cybernetics – Part A: Systems and Humans*, *30*(3), 286–297. https://doi.org/10.1109/3468.844354

Reilly, T. (1996). The St. Petersburg–Tampa Airboat Line: 90 days that changed the world of aviation. *Tampa Bay History*, *18*(2), Article 4.

Rezvani, Z., Jansson, J., & Bodin, J. (2015). Advances in consumer electric vehicle adoption research: A review and research agenda. *Transportation Research Part D: Transport and Environment*, *34*, 122–136. https://doi.org/10.1016/j.trd.2014.10.010

Shively, R. J., Brandt, S. L., Lachter, J., Matessa, M., Sadler, G., & Battiste, H. (2016). Application of Human-Autonomy Teaming (HAT) Patterns to Reduced Crew Operations (RCO). In D. Harris (Ed.), *Engineering Psychology and Cognitive Ergonomics (EPCE 2016)* (Lecture Notes in Computer Science, Vol. 9736, pp. 244–255). Springer, Cham. https://doi.org/10.1007/978-3-319-40030-3_25

Sierzchula, W., Bakker, S., Maat, K., & van Wee, B. (2014). The influence of financial incentives and other socio-economic factors on electric vehicle adoption. *Energy Policy*, *68*, 183–194. https://doi.org/10.1016/j.enpol.2014.01.043

SkyGrid. (2025, April 22). *Autonomy 101: Role of automation in advanced air mobility.* https://www.skygrid.com/autonomy-101-role-of-automation-in-advanced-air-mobility/

Soto-Acosta, P. (2020). COVID-19 pandemic: Shifting digital transformation to a high-speed gear. *Information Systems Management*, *37*(4), 260–266. https://doi.org/10.1080/10580530.2020.1814461

Tuchen, S., Lafrance-Linden, D., Hanley, B., Lu, J., McGovern, S., & Litvack-Winkler, M. (2022). Urban air mobility (UAM) and total mobility innovation framework and analysis case study: Boston area digital twin and economic analysis. *2022 IEEE/AIAA 41st Digital Avionics Systems Conference (DASC)*, 1–14. https://doi.org/10.1109/DASC55683.2022.9925865

Wade, M., & Shan, J. (2020). Covid-19 has accelerated digital transformation, but may have made it harder not easier. *MIS Quarterly Executive*, *19*(3), Article 7.

Wiegmann, D. A., & Shappell, S. A. (2003). *A human error approach to aviation accident analysis: The human factors analysis and classification system.* Ashgate. https://doi.org/10.4324/9781315263878

Wing, D. J., & Cotton, W. B. (2011). *For spacious skies: Self-separation with "Autonomous Flight Rules" in US domestic airspace* (AIAA Paper No. 2011-6865). 11th AIAA Aviation Technology, Integration, and Operations (ATIO) Conference. https://doi.org/10.2514/6.2011-6865

World Economic Forum. (2024). *Advanced air mobility: Shaping the future of aviation.* World Economic Forum.

Yuan, E. S. (2021, March 17; updated October 5, 2022). A year later: Reflecting and looking ahead. *Zoom Blog.*

12 Epilogue

Safety in aviation has always been non-negotiable, but in AAM, it takes on new dimensions. Resilient safety is more than adhering to regulations or preventing accidents—it is a proactive, adaptive mindset that permeates design, operations, and culture. It means building systems that not only avoid failure but also withstand and recover from the unexpected. In the rapidly evolving AAM ecosystem, resilient safety behaves like a living organism: continuously learning, anticipating emerging risks, and strengthening itself over time. This dynamism is pragmatic at its core. It acknowledges that challenges—such as technical malfunctions, human errors, extreme weather events, and cybersecurity threats—will inevitably arise. By anticipating and preparing for these, resilient safety ensures that AAM operations can absorb shocks without catastrophic breakdowns. In doing so, it preserves the continuity of services and the trust of the public, which is the lifeblood of any new transportation system.

Crucially, resilient safety is the bridge between innovation and trust. AAM promises radical new mobility, but that promise means little if people do not feel safe. Thus, resilient safety becomes the *currency of trust*. Each safe flight, each successful automated landing, each well-managed emergency is a deposit into a bank of public confidence. Over time, this trust compounds and pays for the industry's future in very tangible ways, enabling scale and adoption. Safety, therefore, is not a cost or hurdle to innovation; it is the foundation of sustainable innovation. With safety engineered into every aspect of AAM, the industry builds an unassailable reputation.

Before the end of this book, one might wonder what AAM might look like by 2050. A realistic outlook acknowledges that the path will not be linear or easy. Global adversities—economic shocks, climate crises, geopolitical instabilities—are likely to shape the trajectory of AAM. Yet history shows that adversity can *accelerate* innovation and adoption. According to the World Economic Forum (2020), for example, during the COVID-19 pandemic, drones rapidly shifted from niche novelties to "*life-saving tools in the battle against COVID-19*," as the crisis forced society to recognize that their utility outweighed the perceived risks. One industry analysis noted that while the pandemic didn't necessarily accelerate technical invention, "COVID accelerated technology adoption" by changing attitudes and creating urgent needs (Ryan & Divin, 2021). Similarly, the tumultuous decades ahead could catalyze AAM growth. Severe climate events might highlight the value of aerial mobility for disaster relief or evacuation when roads are impassable. Energy shortages or economic pressures could renew emphasis on efficient electric transport networks. Geopolitical tensions might spur investments in decentralized, resilient transportation infrastructure (including AAM) to bolster national supply chain security. Each shock or challenge could push AAM from experimental to essential more quickly than the most optimistic forecasts.

DOI: 10.1201/9781003546306-16

Crucially, it is resilient safety that will make this accelerated adoption possible and prudent. In a future where AAM vehicles are deployed at scale under pressing circumstances, only those operations with mature safety frameworks will earn the public trust and regulatory approval necessary to expand their operations. By 2050, it is likely that AAM will have evolved into a robust component of global transportation, not in spite of global adversities, but partly *because of them*. The network of autonomous electric aircraft ferrying people and goods will have been hardened by trials and refined by lessons learned, much as commercial aviation was shaped by early 20th-century upheavals. AAM services in 2050 are likely to be commonplace in many cities and regions, offering clean, efficient mobility options and emergency response capabilities. They will operate within an ecosystem that has been stress-tested by economic and environmental volatility, and thus built to be fail-safe and adaptable. Ultimately, resilient safety will remain the cornerstone of this industry's endurance, connecting financial stability, operational excellence, and environmental stewardship. Whatever surprises the coming decades bring, a steadfast commitment to resilience in safety will ensure that AAM not only survives turbulent times but also delivers on its promise of sustainable, transformative transportation.

REFERENCES

Ryan, P., & Divin, J. (2021). *COVID-19 – driving changing attitudes to digital asset management*. Aerospace & Defense Review Europe. https://www.aerospacedefensereview.com/cxoinsight/covid19-driving-changing-attitudes-to-digital-asset-management-nwid-733.html

World Economic Forum. (2020, July 6). *COVID-19 is driving the Golden Age of drone delivery*. World Economic Forum. https://www.weforum.org/stories/2020/07/golden-age-drone-delivery-covid-19-coronavirus-pandemic-technology

Annex A: Global eVTOL, eSTOL, and UAS—Active Systems by Continent (October 2025)

North America

Name	Manufacturer (Country)	Type	Primary Purpose	Power Source	Max Range	Max Payload / Capacity	Operational Status
Amazon Prime Air MK30	Amazon (United States)	UAS (delivery drone)	Small package delivery	Electric (battery)	~12 km (7.5 mi) [1]	~2.3 kg (5 lb) payload[1]	Limited service trials (FAA-approved, suburban deliveries)[2]
Midnight	Archer Aviation (United States)	eVTOL (air taxi)	Passenger transport (4 pax + 1 pilot)	Electric (battery)	~160 km (100 mi)[3]	~450 kg (1,000 lb) payload (4 passengers + pilot)[4]	Prototype flight testing; FAA certification in progress (target 2024)[5][3]
ALIA-250	Beta Technologies (United States)	eVTOL (lift+cruise)	Cargo & regional transport (configurable for passengers)	Electric (battery)	~463 km (250 nm)	~635 kg (1,400 lb) or 1 pilot + 5 passengers	Prototype flight trials (cargo demos with UPS/Bristow); certification planned ~2025[6]
EL9	Electra. aero (United States)	eSTOL (hybrid-electric)	Regional passenger & cargo	Hybrid-electric: 4 battery packs + small turbine generator driving 8 wing-mounted electric motors (blown-lift)	330 nm mission with 9 pax/ bags; 1,100 nm ferry (IFR reserves)	9 passengers + pilot (or 3,000 lb cargo)	In development, EL 2 Goldfinch hybrid electric eSTOL demonstrator has flown; EL9 first test flights planned 2027, certification targeted 2029; Evolito selected to supply propulsion units (Oct 2025).)
Chaparral	Elroy Air (United States)	UAS (VTOL cargo drone)	Autonomous cargo delivery (middle-mile)	Hybrid-electric (turbine + battery)[7]	~480 km (300 mi)[8]	~136 kg (300 lb) cargo[8]	Full-scale prototype testing (transition flights achieved); slated for limited logistics ops by 2025[9]

S4 eVTOL	Joby Aviation (United States)	eVTOL (air taxi)	Passenger transport (4 pax + 1 pilot)	Electric (battery)	~240 km (150 mi)[10][11]	~450 kg (1,000 lb) payload (4 passengers + pilot)[12]	Flight testing ongoing (1000+ test flights); FAA certification in progress (service expected ~2025)[10][11]
Hexa	LIFT Aircraft (United States)	eVTOL (multicopter)	Recreational flight (1-person ultralight)	Electric (battery)	~15 km (8–15 mi) per 10–15 min flight[13][14]	1 passenger (max 113 kg/250 lb)[14]	Experimental ultralight—manned demo flights underway; tourist operations planned (no pilot license required)[14][15]
M2	Matternet (United States)	UAS (quadcopter drone)	Medical sample and parcel delivery	Electric (battery)	~20 km (12 mi)[16]	~2 kg (4.5 lb) payload[16]	In service (FAA-certified design[17]; used by UPS and hospitals for routine deliveries)
BlackFly	Opener (United States)	eVTOL (personal ultralight)	Personal air vehicle (1-seat recreational)	Electric (battery)	~64 km (40 mi) (with reserves)[18]	1 operator (max 113 kg/250 lb pilot)[19][20]	In limited production (ultralight category); ongoing flight tests and demonstrations[21][18]
Wing Delivery Drone	Wing (Alphabet) (United States)	UAS (VTOL drone)	On-demand small parcel delivery	Electric (battery)	~19 km (12 mi)[22]	~2.3 kg (5 lb) payload[22]	Active service pilots (operational in the United States, Australia, and Finland, with tens of thousands of deliveries)[23]
Cora / Gen-6	Wisk Aero (United States)	eVTOL (air taxi)	Autonomous passenger transport (4 seats, self-flying)	Electric (battery)	~144 km (90 mi) with reserves[24]	4 passengers (no onboard pilot)[24]	Prototype unveiled (6th-gen); autonomous flight testing ongoing, pursuing FAA certification (mid/late-2020s)[25][24]
P2 Zip	Zipline (United States)	UAS (hybrid fixed-wing)	Medical supply & e-commerce delivery	Electric (battery)	~16 km radius (10 mi) or 38 km one-way[26]	~3.5 kg (8 lb) payload[26]	In service (over 1.4 million deliveries globally)[27]; next-gen P2 system deploying with improved precision tether-drop[28]

(Continued)

Europe

Name	Manufacturer (Country)	Type	Primary Purpose	Power Source	Max Range	Max Payload / Capacity	Operational Status
Black Swan	Dronamics (Bulgaria)	UAS (fixed-wing cargo)	Long-range cargo freight (uncrewed aircraft)	Combustion (jet fuel)	~2,500 km (1,550 mi) [29]	~350 kg (770 lb) payload[29]	Prototype testing (long-range flights in EU/UAE); pursuing certification for cargo operations by 2025[30]
Jet (7-seater)	Lilium GmbH (Germany)	eVTOL (winged jet)	Regional air mobility (air taxi, 6 pax + 1 pilot)	Electric (battery)	~250 km (155 mi) [31]	6 passengers + 1 pilot (or 7-seater autonomous)[32][33]	Flight prototypes in testing (hundreds of flights); certification target ~2026 (company under restructuring in 2024)[31][34]
VX4	Vertical Aerospace (United Kingdom)	eVTOL (vectored thrust)	Urban air taxi (4 pax + 1 pilot)	Electric (battery)	~160 km (100 mi) [35]	4 passengers + 1 pilot (450 kg useful load) [35]	Prototype hovered (2022); ongoing flight tests; EASA certification in progress, service planned ~2026[35] [36]
VoloCity	Volocopter GmbH (Germany)	eVTOL (multicopter)	Urban air taxi (2-seat)	Electric (battery)	~35 km (22 mi)[37]	2 people (including hand luggage, ~200 kg total) [37][38]	Pre-service phase—multiple prototypes flown; targeting EASA type certification by 2024–2025[38][39]
VoloDrone	Volocopter GmbH (Germany)	UAS (heavy-lift drone)	Cargo & utility transport (B2B logistics)	Electric (battery)	~40 km (25 mi) with max load[40]	~200 kg (440 lb) payload[40]	Prototype flight tested (public demo in 2021)[41]; being developed with logistics partners (e.g., DB Schenker) for future use[42][43]
198	Wingcopter GmbH (Germany)	UAS (VTOL drone)	Medical & parcel drone delivery	Electric (battery)	~75 km (47 mi) with 6 kg payload (up to 110 km light load)[44]	~6 kg (13 lb) payload[44]	Limited deployments (trials in Europe, Africa, United States with health and e-commerce projects); production ongoing for global delivery services[45][46]

Asia

Name	Manufacturer (Country)	Type	Primary Purpose	Power Source	Max Range	Max Payload / Capacity	Operational Status
Air One	AIR EV (Israel)	eVTOL (personal)	Personal aerial vehicle (2-seat recreational)	Electric (battery)	~177 km (110 mi) [47]	2 passengers (250 kg combined) [48][49]	Prototype phase (manned test flights started); FAA experimental certificate granted, targeting consumer deliveries in a few years[50][49]
Prosperity I	AutoFlight (China)	eVTOL (lift+cruise)	Air taxi (4 pax + 1 pilot)	Electric (battery)	~250 km (155 mi) [51]	4 passengers + 1 pilot (~350 kg payload)[52][51]	Flight prototypes demonstrated (record 250 km flight[53][54]); CAAC type certification underway (service expected ~2026)[55][51]
FlyCart 30	DJI (China)	UAS (cargo drone)	Short-range freight & drops (winch delivery)	Electric (battery)	16 km with 30 kg load (8 km with 40 kg)[56]	Up to 40 kg (88 lb) payload[57]	Released 2023; being deployed for industrial and emergency logistics (precision drop capable)[58][59]
FlyCart 100	DJI (China)	UAS (cargo drone)	Heavy-lift drone for logistics & EMS	Electric (battery)	12 km with ~65 kg load (6 km with 80 kg max)[60]	Up to 80 kg (176 lb) payload[60]	New 2025 model; entering service for enterprise deliveries (fast swap dual-battery system for <10 min turnaround)[61][62]
EH216	EHang (China)	eVTOL (autonomous)	Urban air taxi (2-passenger drone)	Electric (battery)	~30–35 km (≈22 mi)[63][64]	2 passengers (220 kg total)[63]	Certified—First eVTOL with CAAC type certificate (EH216-S, Oct 2023)[65]; limited tourist and EMS operations initiated in Asia/Middle East[66][67]
SD-05	SkyDrive Inc. (Japan)	eVTOL (flying car)	Air taxi (2-seat short-hop)	Electric (battery)	~15 km (9 mi) initially (aiming 30–40 km with upgrades)[68][69]	2 seats (1 pilot + 1 passenger)[70] [69]	Prototype flight tested publicly (Tokyo, 2023); to be showcased at Expo 2025 Osaka with hoped certification for limited routes[71][68]
Voyager X2	XPeng AeroHT (China)	eVTOL (flying car)	Personal transport (2-seat)	Electric (battery)	~75 km (46 mi) / 35 min per charge[72][73]	2 passengers (~200 kg payload)[74][75]	Prototype demonstrated (uncrewed flights in Dubai and Guangzhou); continuing R&D toward a future drivable flying car model[76][73]

(Continued)

South America

Name	Manufacturer (Country)	Type	Primary Purpose	Power Source	Max Range	Max Payload / Capacity	Operational Status
Eve eVTOL	Eve Air Mobility—Embraer (Brazil)	eVTOL (lift+cruise)	Urban air mobility (4 pax + 1 pilot)	Electric (battery)	~100 km (60 mi) [77][78]	4 passengers + 1 pilot[78]	In development—full-scale prototype unveiled 2023; pursuing ANAC and FAA certification by ~2026[79][77] (significant order backlog for launch in Brazil/US)

Oceania

Name	Manufacturer (Country)	Type	Primary Purpose	Power Source	Max Range	Max Payload / Capacity	Operational Status
Vertiia	AMSL Aero (Australia)	eVTOL (tilt-wing)	Air ambulance & regional transport (5-seat)	Hybrid-electric (hydrogen fuel-cell + battery)[80]	~1,000 km (620 mi) on hydrogen[80]	5 people (4 pax + 1 pilot) or ~500 kg payload [81]	Full-scale prototype in testing (tethered hover achieved)[82][83]; integrating hydrogen powertrain with aim for certification in late-2020s

Sources: Data compiled from official manufacturer specifications, aviation press releases, and industry databases[4][10][56][63][44]. All systems listed are active projects or deployed platforms as of October 2025, with several already in limited service or flight testing. Each entry's range, capacity, and status reflect the latest available information from sources to the best of the author's knowledge, and is provided for informational purposes only (e.g., FAA filings, type certifications, manufacturer announcements)[5][77][65].

Notes:

[1] [2] [22] [23] [26] [27] [28] [29] [30] [56] [57] [58] [59] [60] [61] [62] The Top 10 Cargo Drones in 2025 You Should Know | Grepow. https://www.grepow.com/blog/top-10-cargo-drones-of-2025.html

[3] [4] [5] Archer Aviation—Archer's Airworthiness Criteria for its Midnight Aircraft Published in Federal Register. https://investors.archer.com/news/news-details/2022/Archers-Airworthiness-Criteria-for-its-Midnight-Aircraft-Published-in-Federal-Register/default.aspx

[6] Bristow Signs Order of up to 55 Beta Technologies ALIA-250 eVTOL Aircraft: Bristow Group Inc. (VTOL). https://www.bristowgroup.com/news-media/press-releases/detail/517/bristow-signs-order-of-up-to-55-beta-technologies-alia-250

[7] Kratos to be Exclusive U.S. Manufacturer for the Elroy Air Chaparral. https://www.kratosdefense.com/newsroom/kratos-to-be-exclusive-u-s-manufacturer-for-the-elroy-air-chaparral-vtol-cargo-drone
[8] Chaparral—Aircraft—Elroy Air. https://elroyair.com/chaparral/aircraft
[9] Elroy Air Achieves Major Milestone With Transition of Chaparral. https://elroyair.com/company/news/press-releases/chaparral-VTOL-transitions-to-forward-flight
[10] [11] Joby Becomes First eVTOL Company to Submit Area-Specific Certification Plan: Joby Aviation, Inc. (JOBY). https://ir.jobyaviation.com/news-events/press-releases/detail/32/joby-becomes-first-evtol-company-to-submit-area-specific
[12] Joby Aviation: Joby S4 Aircraft. https://evtol.news/joby-aviation-s4-production-prototype
[13] Army Enlists Personal eVTOL Manufacturer for Casualty Evacuation. https://www.flyingmag.com/army-enlists-personal-evtol-manufacturer-for-casualty-evacuation/
[14] [15] LIFT Aircraft HEXA. https://evtol.news/lift-hexa/
[16] Matternet CEO breaks down drone firm's watershed FAA approval. https://www.freightwaves.com/news/matternet-ceo-breaks-down-drone-firms-watershed-faa-approval
[17] Noise Certification Standards: Matternet Model M2 Aircraft. https://www.transportation.gov/regulations/federal-register-documents/2022-19639
[18] [19] [20] [21] Pivotal BlackFly. https://www.pivotal.aero/
[24] [25] Wisk Unveils World's First Self-Flying, Four-Seat, eVTOL Air Taxi. https://wisk.aero/news/press-release/generation6/
[31] Lilium Seven-Seater eVTOL Jet, Germany—Airport Technology. https://www.airport-technology.com/projects/lilium-7-seater-evtol-jet/
[32] [33] Lilium Jet. *https://lilium.com/*
[34] While Lilium and Volocopter fight for survival, can the remaining. https://tech.eu/2025/02/19/while-lilium-and-volocopter-fight-for-survival-can-the-remaining-players-fill-the-void/
[35] Vertical Aerospace VX4 (prototype)—eVTOL.news. https://evtol.news/vertical-aerospace-VA-1X
[36] Vertical VX4—eVTOL Aircraft for Urban Air Mobility. https://www.aaminternational.com/projects/vertical-vx4/
[37] VoloCity to Become First Commercial Volocopter Aircraft. *https://www.volocopter.com/cn/newsroom/volocity-to-become-first-commercial-volocopter-aircraft*
[38] Volocopter Series D Raises $238 Million—AVweb. http://avweb.com/recent-updates/evtols-urban-mobility/volocopter-series-d-raises-238-million/
[39] Volocopter Could be First to Certify an eVTOL Aircraft. https://www.ainonline.com/aviation-news/advanced-air-mobility/2023-06-21/first-evtol-passenger-flights-set-paris-2024
[40] Volocopter debuts giant drone powerful enough to lift 440 lb. https://newatlas.com/drones/volocopter-volodrone-drone-powerful/
[41] Successful First Public Flight of Volocopter's VoloDrone. https://www.volocopter.com/en/newsroom/successful-public-flight-volodrone
[42] A Trailer for the VoloDrone October 14, 2021—Volocopter. https://www.volocopter.com/en/newsroom/a-trailer-for-volodrone
[43] Volocopter teams up with Near Earth Autonomy to enhance. https://verticalmag.com/news/volocopter-partners-near-earth-autonomy-evtol-volodrone/
[44] Partnership Air Methods—Wingcopter. https://wingcopter.com/air-methods
[45] Wingcopter 198 Wingcopter GmbH—Europe Commercial Drone Guide. https://botsanddrones.net/commercial-drones/f/wingcopter-198

(*Continued*)

[46] Launch Wingcopter 198. https://wingcopter.com/launch-wingcopter-198

[47] Air ONE, Built for fun—Aeroauto. https://aeroautosales.com/air-one/

[48] Personal Air One—AIREv.aero. https://www.airev.aero/air-one

[49] Air EVTOL Air One Price and Review—EV Database. https://ev.motorwatt.com/ev-database/flying-electric-vehicles/air-evtol-air-one

[50] FAA Grants Experimental Airworthiness Certification for Air One. https://www.ainonline.com/aviation-news/futureflight/2025-09-02/faa-grants-experimental-airworthiness-certification-air-one

[51] AutoFlight Begins Chinese Certification of Prosperity eVTOL Aircraft. https://www.ainonline.com/aviation-news/futureflight/2024-04-30/autoflight-begins-chinese-certification-prosperity-evtol

[52] Prosperity I—AutoFlight. https://autoflight.com/assets/uploads/files/autoflight-prosperity-i-specs-062023.pdf

[53] [54] [55] China eVTOL Aircraft Round-Up. https://evtol.news/news/china-evtol-aircraft-round-up

[63] [65] EHang eyes 2023 CAAC certification; civilian flights underway. https://www.flightglobal.com/aerospace/ehang-eyes-2023-caac-certification-civilian-flights-underway/153193.article

[64] China's EHang Earns World's First eVTOL Type Certificate. https://www.flyingmag.com/chinas-ehang-earns-worlds-first-evtol-type-certificate/

[66] [67] EH216-S Pilotless eVTOL Completes Debut Passenger Flights in. https://ir.ehang.com/news-releases/news-release-details/eh216-s-pilotless-evtol-completes-debut-passenger-flights

[68] SkyDrive SD-05 eVTOL Showcased at Osaka Kansai Expo—Dronelife. https://dronelife.com/2025/04/10/skydrive-showcases-public-flight-of-sd-05-evtol-at-osaka-kansai-expo/

[69] Japan's SkyDrive sees market for lighter, short-range air taxis. https://verticalmag.com/news/japans-skydrive-sees-market-for-lighter-short-range-air-taxis/

[70] SkyDrive Unveils SD-05 Flying Car Design. https://en.skydrive2020.com/archives/7418

[71] SkyDrive Conducts Public Flight of SKYDRIVE SD-05 during Media. https://en.skydrive2020.com/archives/15064

[72] Australia's First Flying Car Takes Off—AfMA. https://afma.org.au/australias-first-flying-car-takes-off/

[73] XPeng X2 takes flight in Australia—Autoblog. https://www.autoblog.com/news/xpeng-x2-takes-flight-in-australia

[74] [75] XPENG X2—a futuristic two-seater eVTOL flying car—Aeroauto. https://aeroautosales.com/xpeng-x2-a-two-seater-evtol-flying-car/

[76] This electric flying car drives like a 'normal car' at speeds of up to 80 mph. https://interestingengineering.com/transportation/electric-flying-car-drives-80mph

[77] A look at Eve's eVTOL aircraft as it targets type certification in late 2025—Vertical Mag. https://verticalmag.com/features/eve-evtol-aircraft-design-approach/

[78] Eve Air Mobility. https://www.eveairmobility.com/

[79] Brazil's ANAC sees realistic path to certify Embraer's Eve eVTOL by late 2027. https://businessaviation.aero/evtol-news-and-electric-aircraft-news/low-altitude-economy/brazils-anac-sees-realistic-path-to-certify-embraers-eve-evtol-by-late-2027

[80] [81] Vertiia | AMSL Aero. https://www.amslaero.com/our-product

[82] [83] AMSL Aero Prepares to Test Hydrogen-powered Vertiia eVTOL Aircraft | Aviation International News. https://www.ainonline.com/aviation-news/futureflight/2025-03-25/amsl-prepares-test-hydrogen-powered-vertiia-evtol-aircraft

Annex B: Notable UAS-Related Accidents and Incidents (January 2016–October 2025)

Date	Location	Category	Incident Description	Outcome/Severity	References
19 Feb 2016	Canberra, ACT, Australia	Other incident	DJI Phantom drone crashed during a public ceremony at the Australian War Memorial (within a no-fly zone). It landed near VIP attendees (including a former Defense Minister).	No injuries. The incident raised security concerns. CASA (Civil Aviation Authority) investigated and fined the recreational pilot $900 for flying too close to people and in restricted airspace (closing the matter) [60][61]. Highlighted the need for geofencing of sensitive sites.	Flight Safety Australia, 2016 [60][61]
19 Feb 2016	Paris (CDG vicinity), France	Near-miss	Air France Airbus A320 (from Barcelona) nearly hit a drone at ~1,600 m (5,200 ft) on approach to Charles de Gaulle Airport [44]. Pilot spotted the drone ~5 m away and had to disengage autopilot to avoid it.	Critical near-miss—the drone passed ~5 meters from the jet. The pilot took evasive action. French BEA opened its first drone-related airprox investigation [44]. The incident was rated as a serious risk. Prompted calls in France for stricter drone exclusion near airports.	Bureau d'Enquêtes et d'Analyses [BEA], 2016; Aviation24.be, 2016; [44].
10 Mar 2016	Yorkshire, England, United Kingdom	Other incident	Drone vs. train: A hobby drone collided with the historic "Flying Scotsman" steam locomotive during a celebratory run [65][66]. The drone was being flown illegally close to the railway.	Property damage—minor (the train was not derailed; the extent of damage to the locomotive was minimal). British Transport Police identified the drone operator and issued warnings/citations [65][66]. This incident (drone vs train) highlighted that drones can endanger more than just aircraft. BTP reminded the public that it's illegal to fly within 50 m of a train [65][67].	British Transport Police via ITV News, 2016; The Telegraph, 2016; [65][66]
18 Mar 2016	Los Angeles, California, United States	Near-miss	Lufthansa Airbus A380 jumbo jet (LAX-bound) encountered a drone at ~5,000 ft, about 14 miles east of LAX [55]. The drone was reported to have passed ~200 ft above the A380's flight path.	Serious near-miss—no collision. The FBI and FAA investigated [55], marking one of the earliest major near-miss reports in U.S. airspace. Prompted FAA warnings that a drone ingested in an engine could cause a real threat (Williams, FAA UAS office) and led to renewed "If you fly, we can't" campaigns during wildfire season.	Serna, 2016 (LA Times)[55].

17 Apr 2016	London, England, United Kingdom	Collision (alleged)	British Airways Airbus A320 on approach to Heathrow reported hitting a drone at ~1,700 ft. Incident made headlines amid growing drone fears [31][32].	No damage found on airliner upon inspection; no drone debris found [33][34]. U.K. Transport Secretary later stated experts concluded it was "not a drone incident" (possibly a plastic bag or other object) [33][34]. AAIB investigation closed due to lack of evidence. Led to calls for stricter drone rules (registration, geofencing) [35][36].	Press Assoc., 2016[33][34]; McLoughlin, 2016.
25 Apr 2016	Kent, England, United Kingdom	Other incident	Prison contraband delivery: A man used a drone to drop drugs and mobile phones into prisons (HMP Swaleside and others) at night [68][69]. He was caught by police with a spray-painted drone in his car trunk.	Criminal case—the perpetrator was convicted, marking the United Kingdom's first drone smuggling conviction [68][69]. He received a prison sentence; an accomplice was cleared. The case underscored the security threat from drones and led to improved prison drone defenses (e.g., netting, RF jammers).	[68][69].
5 June 2016	Beloeil, Quebec, Canada	Other incident	DJI Phantom 3 drone filming a 5K run fell onto a spectator (38-year-old woman) in the crowd [62]. She was struck on the head. Another videographer had reportedly just warned the Phantom's operator about flying over people [63].	Injury accident—the woman suffered a concussion and whiplash; taken to the hospital [62]. She recovered, but missed work for weeks. The drone operator claimed to be perplexed by the crash; the TSB opened an investigation [64]. He was later fined for careless operation (flying over a crowd). The incident emphasized the danger of drone flights over people.	[62][63].
31 Jul 2016	Edmonton, Alberta, Canada	Near-miss	Police helicopter (Edmonton Police Air-1) encountered a white quadcopter drone at ~750 ft while responding to a call downtown at night [42]. The drone approached the heli, coming dangerously close.	Serious near-miss—Air-1 crew reported a "catastrophic" outcome was likely avoided only by chance. No collision or damage. A criminal investigation was opened for endangering an aircraft [42]. Federal (Transport Canada) charges were also considered.	[42].

(*Continued*)

Date	Location	Category	Incident Description	Outcome/Severity	References
16 Aug 2016	Near Klagenfurt, Austria	Near-miss	Rescue helicopter (air ambulance) transporting a patient nearly collided with a small quadcopter drone that came within meters during cruise [41]. The pilot had to perform an evasive maneuver.	High risk near-miss—no collision occurred. The drone's operator was not identified by local police [41]. The incident spurred public warnings in Austria about drone flights near emergency aircraft.	[41].
26 Aug 2016	London, England, United Kingdom	Near-miss	Eurocopter EC145 air ambulance (East Anglian Air Ambulance) nearly collided with a drone over north London (7:45 pm) at ~1000 ft [46]. Crew (not including Prince William, who often flies this unit) saw the drone pass ~30 m away.	High-risk near-miss—UK Airprox Board classified it Category A (serious risk of collision) [47]. The drone was flying over a populated area without CAA permission, violating regulations [46]. Incident highlighted dangers even to emergency helicopters; prompted renewed warnings to drone users.	[46][48].
15 Jan 2017	Hangzhou, Zhejiang, China	Near-miss	Multiple airliners on approach to Xiaoshan Intl Airport were overflown by a DJI Mavic drone, as evidenced by an onboard video posted online [43]. One clip showed a passenger jet landing with a drone in proximity.	Illegal drone operation—no collision, but risk was high at ~450 m (1480 ft) altitude. The 23-year-old drone pilot was detained by Chinese authorities once the video surfaced [43]. DJI condemned the incident. Led to increased enforcement of no-drone zones near airports in China.	[43].
19 May 2017	Edinburgh, Scotland, United Kingdom	Near-miss	Loganair Saab 2000 (regional airliner) descending into Edinburgh (from Shetland) had to take evasive action to avoid a drone at ~4,000 ft altitude [51]. The drone passed ~20–30 m from the aircraft.	High-risk near-miss—successful avoidance maneuver, no collision. Police traced the drone's probable launch area (~Bathgate, 20 km away) [51][52]. The incident spurred Scottish authorities to educate the public on drone restrictions (no flying near approach paths).	[51][52].
2 July 2017	Gatwick Airport, United Kingdom	Other incident	Runway closure (brief): Gatwick Airport operations were halted twice (9 minutes and 5 minutes) due to a drone spotted near the runway [70][71]. Several flights were diverted as a precaution.	Disruption—minor delays and 5 flight diversions (4 EasyJet, 1 BA) [72]. Sussex Police investigated [72]; the drone operator wasn't found. This previewed the far more serious Gatwick incident in 2018, and prompted U.K. airports to review their emergency procedures for drone sightings.	[70][71][72]

9 July 2017	Gatwick (near), England, United Kingdom	Near-miss	British Airbus A319 (130 passengers) on approach to Gatwick had a large quadcopter drone fly over its right wing at ~100 m altitude on final approach (20:35 BST)[3]. Initially thought to be a bird until it came very close to the wingtip and fuselage.	Extremely high risk—pilots could not evade; the drone (≈1 m size) came within a few meters [4]. The UK Airprox Board report concluded only luck prevented a disaster, saying a larger aircraft "*might not have missed it*" [3]. Police were notified, but the operator was not found. Intensified calls for drone registration deadlines [53].	[4][54]
21 Sep 2017	Staten Island, New York, United States	Collision (verified)	DJI Phantom 4 hobby drone vs. U.S. Army UH-60M Black Hawk helicopter (operating security patrol for UN General Assembly)—mid-air impact at ~300 ft altitude [1].	Minor damage to the helicopter (main rotor blade, window frame) but landed safely; drone destroyed; no injuries. NTSB investigation found drone pilot at fault for flying beyond line-of-sight and into restricted airspace [7][8].	[1][7]
12 Oct 2017	Quebec City, Quebec, Canada	Collision (verified)	Unknown drone vs. Beechcraft King Air A100 (Skyjet) on approach to Jean Lesage Intl Airport [9]—first confirmed drone strike on a passenger aircraft in North America.	Minor damage to wing (scratches, paint transfer) [2]; crew declared emergency, landed safely; no injuries. Drone was flying at ~1500 ft (5× legal altitude); operator not found. Transport Canada and TSB investigated (interim safety rules were issued) [9][10].	[2][10].
25 Mar 2018	Auckland, New Zealand	Near-miss	Air New Zealand Boeing 777-200 on final approach (landing gear down) encountered a drone that came within an estimated 5 m of the aircraft at ~500 ft altitude [45]. Pilots were unable to take evasive action in time.	Severe near-miss—no impact, but crew feared the drone could be ingested into an engine [45]. The offending UAV operator remains unknown. The incident triggered public outcry; Air New Zealand called for tougher drone laws and even prison terms for reckless operators [45].	[45]

(*Continued*)

Date	Location	Category	Incident Description	Outcome/Severity	References
4 June 2018	Johnson Valley, California, United States	Collision (verified)	DJI Mavic 2 Zoom drone vs. Airbus AS350BA news/film helicopter (covering an off-road race)—drone struck the helicopter's tail rotor during low-altitude operations [11][12].	Significant damage: tail rotor blades bent, causing severe vibration [12]. Pilot executed a safe emergency landing; no injuries. Helicopter was grounded and removed by trailer. NTSB found the drone pilot failed to yield right-of-way and lacked coordination with manned aircraft (probable cause: drone pilot's improper operation) [13][14].	[13][14].
10 Aug 2018	Driggs, Idaho, United States	Collision (verified)	DJI Mavic Pro drone vs. Hot-air balloon (sightseeing flight near Driggs Airport)—*first-ever recorded* drone–balloon mid-air collision [15]. The drone pilot (hobbyist) lost sight of the balloon while filming it.	No damage to balloon (drone's rotors sheared off on the balloon's envelope/lines); balloon landed safely; no injuries. Drone destroyed. FAA rules were violated (flight within 5 miles of the airport without notice), and pilot error was cited [16][17]. Incident reported to NTSB for awareness.	[16][17].
Nov 2018	Las Vegas, Nevada, United States	Near-miss	Frontier Airlines Airbus A320 on final approach to Las Vegas McCarran had a drone hover *above* it on the glide path—an incident captured in a drone's own YouTube video that surfaced online [56]. The video showed the airliner passing underneath the hovering quadcopter.	Reckless near-miss—no contact, but the video evidence sparked outrage. Aviation organizations (AMA, AUVSI) condemned the drone operator's actions, and the FAA launched an investigation [57]. This highlighted the need for better geofencing and enforcement, as the drone pilot was never publicly identified as of 2018.	[56][57]
13 Dec 2018	Tijuana, Baja California, MX	Collision (alleged)	Aeroméxico Boeing 737-800 (Flight 773) on approach to Tijuana Airport reported a loud bang. After landing, the nose radome was found dented [37][38]. The crew suspected a drone strike in flight.	Aircraft surface damage (radome dent). Landed safely; no injuries. Investigation treated it as a possible drone strike [37], but the cause was not conclusively determined (no drone found). Incident under investigation by the airline and authorities as of 2018 [39][40].	[39][40].

19–21 Dec 2018	Gatwick Airport, United Kingdom	Other incident	Major airport disruption: Repeated drone sightings (or reports thereof) over Gatwick's runway led to complete airport closure for ~45 hours [5][6] during the busy pre-Christmas period. Every time the airport was prepared to reopen, new sightings were reported.	Severe disruption—about 1,000 flights were canceled or diverted, affecting ~140,000 passengers [5][6]. Army anti-drone units were deployed. Two suspects were arrested but released with no charges [73]. Ultimately, no physical evidence of any drone was found [74], leading some experts to doubt a drone was ever present. The incident exposed serious gaps in drone detection and caused worldwide alarm in the aviation community.	[6][75].
8 Jan 2019	Heathrow Airport, United Kingdom	Other incident	Temporary flight halt: Departures at Heathrow were suspended for about an hour in the evening after a drone sighting was reported near the airfield [76][6]. Coming just weeks after Gatwick's chaos, this triggered a swift response.	Moderate disruption—several flights held on ground; no long-term cancellations. Military anti-drone systems were activated as a precaution. No drone was found or confirmed. Incident added pressure for U.K. authorities to expedite airport drone-defense measures (Heathrow installed radar/UAVs detection post-event).	[6][76]
4 June 2019	Manassas, Virginia	Other incident	Aurora Flight Sciences PAV (Pegasus) VTOL test, eVTOL / AAM prototype—vibration triggered false ground mode, commanded idle; hard landing.	No injuries; Substantial damage.	Aviation Safety Network, 2019a [91]
21 June 2019	Beeville, Texas	Other incident	Boeing CV2 Cargo Air Vehicle (Cargo UAS prototype) deviated in higher-than-expected crosswind; geo-fence contingency autoland/cut-power landing.	No injuries; Substantial damage.	Aviation Safety Network, 2019b [92]

(*Continued*)

Date	Location	Category	Incident Description	Outcome/Severity	References
Dec 2019–Jan 2020	Eastern Colorado & Nebraska, United States	Other incident	Unidentified drone swarms: Residents and law enforcement reported nightly formations of large drones (6-foot wingspans) flying grid patterns over rural Colorado and Nebraska [78]. Sightings of groups of 6–10 drones sparked mystery and concern. The FAA and FBI launched investigations.	Enforcement challenge—despite dozens of sightings, authorities could not confirm the drones' origins [79][80]. Speculation ranged from hobby groups to military exercises, but no operator was found. The phenomena stopped as abruptly as they began. This incident highlighted limitations in tracking low-flying drones over wide areas and led to task forces to improve drone surveillance and reporting (no safety incidents occurred, aside from one near-miss with a medical helicopter) [81][82].	[81][82].
25 July 2020	Portland, Oregon, United States	Other incident	Drone in restricted airspace: During nightly protests in Portland, a man was arrested for flying a drone in restricted downtown airspace near a federal courthouse. The city was under FAA temporary flight restrictions (TFR) at the time due to civil unrest.	Legal action—the 33-year-old operator was federally charged with unsafe operation of an unmanned aircraft in restricted airspace [83][84]. This was one of the first such U.S. prosecutions. The incident underscored that TFR violations by drones (even if no injury resulted) carry serious consequences. It also demonstrated drones being used amidst protests, raising new security issues.	[83][84]
10 Aug 2021	Markham, Ontario, Canada	Collision (verified)	DJI Matrice 210 police drone (York Regional Police) vs. Cessna 172N trainer (Canadian Flyers Intl) on final approach to Buttonville Airport [18][19]. The drone was hovering at ~400 ft AGL for a police operation.	Major damage to Cessna: front left engine cowling dented and propeller strike observed [19]. Aircraft landed safely; no injuries. Drone destroyed. TSB investigation found the student pilots were unaware of the drone; inadequate visual scanning and lack of effective "see-and-avoid" by both parties contributed (visual observer not properly briefed) [20][21].	[20][21]

22 Jan 2022	Brentford, London, United Kingdom	Other incident	Sports event disruption: A drone overflew a Premier League football match (Brentford vs. Wolves) at Brentford Community Stadium, London. Referees halted the game for ~20 minutes due to safety concerns [77].	Public safety risk—the match was suspended until the drone left. No injuries or damage, but it demonstrated how drones can disrupt large events. Police investigated; the operator was not immediately identified. The FA and stadium authorities reiterated that drones are banned over football crowds for safety.	[77]
20 Feb 2024	Orlando, Florida	Other incident	Apellix B1 washing drone experienced magnetometer issue; operator grabbed UAS and was injured.	One serious injury.	National Transportation Safety Board, 2025a [93]
Nov–Dec 2024	New Jersey & New York, United States	Other incident	Mystery large drones: Over several weeks in late 2024, hundreds of residents (and even the NJ Governor) reported sightings of multiple large drones at night across New Jersey, New York City outskirts, and neighboring states [85][86]. Drones were described as bright, low-flying objects sometimes near critical infrastructure. Federal agencies (FBI, DHS, FAA) were engaged.	Public concern—despite extensive investigations, officials could not verify a source or threat. The drones did not interfere with civil aviation beyond entering some restricted areas (e.g., over a Navy weapons station, where they were detected but not intercepted) [87][88]. Authorities stated there was no evidence of malicious intent or foreign control [89]. The wave of sightings subsided without incident. This episode highlighted gaps in drone identification capabilities and fueled calls for better radar tracking of UAS at night.	[87][90]
21 Dec 2024	Orlando, FL, United States	Accident (multi-UAS)	Drone-show swarm of Uvify IFO sUAS: launch parameter file mis-sent & 7° show rotation → early collisions and falling drones; child seriously injured.	One serious injury (spectator); numerous sUAS damaged.	National Transportation Safety Board, 2025b [94]

(Continued)

Date	Location	Category	Incident Description	Outcome/Severity	References
9 Jan 2025	Los Angeles, California, United States	Collision (verified)	Unidentified quadcopter drone vs. Canadair CL-415 "Super Scooper" firefighting airplane (on wildfire water-drop mission during Palisades Fire). Drone illegally flew into an active fire TFR and struck the waterbomber's wing [26][27].	Moderate damage: 3×6 inch hole torn in left wing [26]. Aircraft landed safely; no injuries, but taken out of service ~5 days for repairs [28]. The drone was destroyed. A 56-year-old drone operator was later identified and pled guilty to unsafe operation (facing fines >$65k and up to 1 year in prison) [29][30].	[30][26].
17 Feb 2025	Philadelphia, Pennsylvania	Collision (verified)	Hospital EC135T2 helicopter struck DJI Air 3 during approach; rotor blade leading-edge damage.	No injuries; Substantial damage to helicopter rotor.	National Transportation Safety Board, 2025c [95]
21 Feb 2025	Pendleton, Oregon	Near-collission	Amazon MK30 SCL test: detected an overhead drone, descended into obstacle (cart), then unplanned ascent and crash.	No injuries; Substantial damage.	National Transportation Safety Board, 2025d [96]
28 Mar 2025	Dixon, California	Other incident	Pyka Pelican test flight; transition/abort sequence led to hard landing and structural damage.	No injuries; Substantial damage.	National Transportation Safety Board, 2025e [97]
7 July 2025	Kerrville, Texas, United States	Collision (verified)	DJI Mavic 3 Enterprise drone (Texas state police) vs. UH-60 Black Hawk helicopter (Texas Army Nat Guard)—mid-air collision during flood search and rescue (SAR) operations (temporary restricted airspace) [22][23]. Initially misidentified as a civilian rogue drone.	Negligible damage to Black Hawk (no immediate damage evident); landed safely; no injuries. Helicopter grounded ~2 hours for inspection. Incident occurred *within* permitted drone mission—investigation found the official drone ascended too high and stalled into the heli's flight path [24][25]. Local authorities corrected earlier claims that an illegal drone was to blame.	[24][25]
16 Sept 2025	Changchun, Jilin (China)	Collision (verified)	Two XPeng AeroHT "flying cars" collided during air-show rehearsal; one caught fire; one pilot injured.	1 injured (non-life-threatening).	Rahman-Jones (2025) [98]

Notes:

"No injuries" indicates no human injuries were reported in the incident. Altitudes are given in local units (feet or meters) as reported. All drone operations described violated some aspect of existing regulations (e.g., beyond line-of-sight, above altitude limits, in restricted areas, or over people) – reflecting the regulatory challenges posed by rogue drones. Each entry reflects the latest available information from sources to the best of the author's knowledge and is provided for informational purposes only

[1] [7] [8] NTSB Finds Recreational Drone Operator Caused Crash with Army BlackHawk Helicopter—Dunning Vallejo & MacDonald LLP https://dvmlaw.com/news/ntsb-finds-recreational-drone-operator-caused-crash-with-army-blackhawk-helicopter/

[2] [9] [10] Air transportation safety investigation A17Q0162—Transportation Safety Board of Canada. https://www.tsb.gc.ca/eng/enquetes-investigations/aviation/2017/a17q0162/a17q0162.html

[3] [4] [5] [6] [15] [16] [17] [18] [19] [22] [23] [31] [32] [35] [36] [37] [38] [39] [40] [41] [42] [43] [44] [45] [46] [47] [48] [49] [50] [51] [52] [53] [54] [55] [56] [57] [58] [59] [61] [62] [63] [64] [65] [66] [67] [68] [69] [70] [71] [72] [73] [74] [75] [76] [77] [78] List of unmanned aerial vehicle-related incidents—Wikipedia. https://en.wikipedia.org/wiki/List_of_unmanned_aerial_vehicle-related_incidents

[11] [12] Drone hits helicopter in Baja 500 | San Diego Reader. https://www.sandiegoreader.com/news/2018/jun/08/stringers-drone-hits-helicopter-baja-500/

[13] [14] DJI Mavic 2 Drone Pilot to Blame for Collision With Helicopter Reports NTSB. https://dronexl.co/2022/02/16/dji-mavic-2-drone-collision-helicopter-ntsb/

[20] [21] TSB releases findings of probe into 2021 aircraft, YRP drone collision at Buttonville Airport—Toronto | Globalnews.ca. https://globalnews.ca/news/9421883/tsb-buttonville-yrp-drone-crash-2021/

[24] [25] Kerrville clarifies drone involved in helicopter incident was authorized. https://news4sanantonio com/news/local/kerrville-clarifies-drone-involved-in-helicopter-incident-was-authorized-flooding-hunt-camp-mystic-search-and-rescue

[26] [27] [28] [29] [30] Drone operator charged for hitting 'super scooper' plane during Palisades Fire: DOJ—*ABC News*. https://abcnews.go.com/US/drone-operator-charged-hitting-super-scooper-plane-palisades/story?id=118313936

[33] [34] Heathrow plane strike "not a drone incident" | Drones (non-military) | *The Guardian*. https://www.theguardian.com/technology/2016/apr/28/heathrow-ba-plane-strike-not-a-drone-incident

[60] Flight Safety Australia. (2016, February 22). *Drone crash at Australian War Memorial under investigation*. Civil Aviation Safety Authority (CASA). https://www.flightsafetyaustralia.com/2016/02/drone-crash-at-australian-war-memorial-under-investigation/

[79] [80] [81] [82] Mystery drone investigation intensifies after close encounter with medical helicopter—*ABC News*. https://abcnews.go.com/US/mystery-drone-investigation-intensifies-close-encounter-medical-helicopter/story?id=68201092

[83] [84] District of Oregon | 22 Arrested, Facing Federal Charges After Weekend Protest at Federal Courthouse in Portland | United States Department of Justice. https://www.justice.gov/usao-or/pr/22-arrested-facing-federal-charges-after-weekend-protest-federal-courthouse-portland

(*Continued*)

[85] [86] [87] [88] [89] [90] Why Are There So Many Unexplained Drones Flying Over the United States? https://www.csis.org/analysis/why-are-there-so-many-unexplained-drones-flying-over-united-states

[91] Aviation Safety Network. (2019a, June 4). *Accident Boeing/Aurora Flight Sciences Pegasus N83AU, Tuesday 4 June 2019*. https://aviation-safety.net/wikibase/226471

[92] Aviation Safety Network. (2019b, June 21). *Accident Boeing CV2 N243XT, Friday 21 June 2019*. https://aviation-safety.net/wikibase/227103

[93] National Transportation Safety Board. (2025a). *Aviation Investigation Final Report ERA24LA216: Apellix B1 Washing Drone, FA33N7WLTR, Orlando, Florida, February 20, 2024*. https://data.ntsb.gov/carol-repgen/api/Aviation/ReportMain/GenerateNewestReport/194250/pdf

[94] National Transportation Safety Board. (2025b, January 16). *Aviation Investigation Preliminary Report DCA25LA065: Uvify IFO swarm, Lake Eola, Orlando, FL, December 21, 2024* [PDF]. https://www.newsomelaw.com/wp-content/uploads/2025/01/NTSB-Preliminary-Report-DCA25LA065-1.16.25.pdf

[95] National Transportation Safety Board. (2025c). *Docket ERA25LA129: Aviation Investigation, Philadelphia, PA, February 17, 2025* (Docket No. ProjectID=199768). https://data.ntsb.gov/Docket?ProjectID=19976

[96] National Transportation Safety Board. (2025d). *Docket WPR25LA103: Aviation Investigation, Pendleton, OR, February 21, 2025* (Docket No. ProjectID=199746). https://data.ntsb.gov/Docket?ProjectID=19974

[97] National Transportation Safety Board. (2025). *Docket DCA25LA174: Aviation Investigation, Dixon, CA, March 28, 2025* (Docket No. ProjectID=199962). https://data.ntsb.gov/Docket?ProjectID=199962

[98] Rahman-Jones, I. (2025, September 17). Flying cars crash into each other at Chinese air show. *BBC News*. https://www.bbc.com/news/articles/cn5qk4de997o

Annex C: AAM Hazard Identification and Risk Assessment (HIRA) Illustrative Examples

Likelihood → Severity ↓	Rare (1)	Unlikely (2)	Possible (3)	Likely (4)	Almost certain (5)
Catastrophic (5)	Review	Review	Unacceptable	Unacceptable	Unacceptable
Major (4)	Acceptable	Review	Review	Unacceptable	Unacceptable
Moderate (3)	Acceptable	Acceptable	Review	Review	Unacceptable
Minor (2)	Acceptable	Acceptable	Acceptable	Review	Review
Insignificant (1)	Acceptable	Acceptable	Acceptable	Acceptable	Acceptable

FIGURE C.1 Risk heatmap.

TABLE C.1
Examples of AAM HIRA

ID	Hazard	L (Initial)	S (Initial)	Initial Level	Mitigations	L (Residual)	S (Residual)	Residual Level	Tolerable (Residual)	ALARP (Residual)
A1	Critical aircraft system failure (C2/GPS/propulsion)	2	5	Review	DAL-A [1] design & segregation; robust C2; BIT; safe-degrade & contingency procedures; rigorous maintenance	1	5	Review	Yes	Yes
A2	Disparate social & environmental impacts	3	3	Review	Equity/EJ analysis; community benefits agreements; service subsidies; careful siting; transparent metrics	2	3	Acceptable	Yes	Yes
A3	Loss of aircraft control (flight-control failure)	2	5	Review	Redundant controls/sensors; envelope protection; degraded modes; testing/simulation; operator training	1	5	Review	Yes	Yes
A4	Autonomy / complex software failure	3	5	Unacceptable	DO-178C processes; formal verification; independent V&V; runtime monitors/guards; controlled OTA updates	1	5	Review	Yes	Yes
A5	Flight outside approved airspace	3	4	Review	Hard geofencing; UTM/USS conformance monitoring; Remote ID/ADS-B; ATC coordination; training	1	4	Acceptable	Yes	Yes
A6	Unaffordability of services	4	2	Review	PPPs; demand aggregation; tiered pricing; scale efficiencies; prioritize essential services	3	2	Acceptable	Yes	Yes
A7	Privacy breach (unauthorized use/sharing)	3	3	Review	Privacy-by-design; data minimization; encryption; strong IAM & audit; consent/opt-out; incident response	2	3	Acceptable	Yes	Yes

A8	Privacy concerns from low-altitude flight	4	2	Review	Published corridors/altitudes; avoid sensitive areas; limit non-nav sensors; engagement & complaint SLAs	2	2	Acceptable	Yes	Yes
A9	Unsafe proximity to people/property	3	5	Unacceptable	Detect-and-avoid; strategic separation & keep-out volumes; minimum altitudes; emergency descent/parachute; robust CONOPS	1	5	Review	Yes	Yes
A10	Scaled operational noise (network-level)	4	4	Unacceptable	Noise budgets/caps; vertiport siting & profiles; quiet hours; fleet noise certification/monitoring; slotting/demand management	2	4	Review	Yes	Yes
A11	Terrorism/sabotage/insider threat	2	5	Review	Vetting & insider-threat program; secure facilities; screening; tamper-evident design; restricted-site geofencing; LE coordination	1	5	Review	Yes	Yes
A12	Visual pollution	3	2	Acceptable	Route/altitude planning; design/livery guidelines; caps near sensitive/scenic areas; public trials & engagement	2	2	Acceptable	Yes	Yes
A13	Ground incidents (FOD, HV shock, vehicle strikes, fires)	3	4	Review	FOD walks; ramp layout & marshaling; HV LOTO & PPE; battery safety & fire suppression; vehicle speed limits; training	2	4	Review	Yes	Yes
A14	Overwhelming ATC in transitions/mixed airspace	3	4	Review	UTM–ATC integration; corridors/strata; equipage requirements; capacity modeling; traffic caps/staffing plans	2	4	Review	Yes	Yes

(Continued)

TABLE C.1 (*Continued*)
Examples of AAM HIRA

ID	Hazard	L (Initial)	S (Initial)	Initial Level	Mitigations	L (Residual)	S (Residual)	Residual Level	Tolerable (Residual)	ALARP (Residual)
A15	Human error (task saturation, miscommunication, loss of SA)	4	4	Unacceptable	HF-optimized HMI; SOPs & checklists; CRM/SRM; FOQA/LOSA; fatigue limits; remote-ops staffing ratios & alerting aids; recurrent training	2	4	Review	Yes	Yes
A16	Passenger incidents (unruly/unsafe/disruptive)	3	3	Review	Passenger policies/screening; automated cabin monitoring; safety briefings/restraints; escalation/diversion procedures	2	3	Acceptable	Yes	Yes
A17	Cybersecurity compromise (e.g., hacking)	3	5	Unacceptable	DO-326A/ED-202A; threat modeling; segmentation; secure boot & code signing; supply-chain security	2	5	Review	Yes	Yes
A18	Airspace congestion	4	3	Review	UTM strategic deconfliction; slots/metering; dynamic rerouting; performance-based separation; demand management	2	3	Acceptable	Yes	Yes
A19	Single-aircraft noise nuisance	3	3	Review	Low-noise propulsors/tip speeds; trajectory smoothing; higher altitudes; noise-abatement arrivals/departures; community limits	2	2	Acceptable	Yes	Yes

NOTE

1. DAL-A = the highest design assurance level in aerospace standards (e.g., DO-178C for software, DO-254 for airborne electronics, and ARP4754A at the system level).

Glossary

AAM Implementation (United States): FAA's Innovate28 (I28) implementation plan outlines how AAM operations scale into the NAS (initially piloted) and how agencies coordinate on procedures, airspace, and infrastructure.

ACAS X (ACAS Xu/Xr): The next-gen airborne collision avoidance family; ACAS Xu is tailored for UAS and complements DAA to reduce mid-air collision risk (RTCA DO-386).

Advanced Air Mobility (AAM): An emerging air transport system that moves people and goods using new aircraft (often highly automated and electrically powered), integrated into the NAS for local, regional, and public service missions. NASA frames AAM as "*bringing the movement of people and goods off the ground, on demand, and into the sky*"; FAA's AIM also describes AAM as a sector focused on safely integrating highly automated aircraft.

ARC: Baseline Air Risk Class

Detect-and-Avoid (DAA): Functions that enable UAS/piloted aircraft to remain well clear and avoid collisions. RTCA DO-365 defines DAA MOPS and the *remain-well-clear* concept used in U.S. UAS integration.

Digital twin: A virtual replica of a physical object, system, or process that mirrors its real-world counterpart using real-time data from sensors and interconnected devices.

Environmental qualifications (DO-160): Standard test procedures and environmental categories for airborne equipment (temperature, vibration, RF/EMC, etc.) used by certification authorities worldwide.

eSTOL: Electric short takeoff and landing aircraft (fixed-wing) that use distributed electric propulsion and blown-lift/high-lift systems to operate from very short runways, offering eVTOL-like mission access with lower energy per operation.

eVTOL: Electric VTOL aircraft using electric propulsion for vertical takeoff/landing and cruise (full-electric or hybrid-electric).

Fail-operational vs. fail-safe (VTOL context): In EASA SC-VTOL terms, continued safe flight and landing (Enhanced category) aligns with fail-operational behavior; controlled emergency landing (Basic category) aligns with fail-safe behavior.

FIMS (Flight Information Management System): The FAA-managed interface for data exchange between FAA systems and UTM participants—central to UTM pilots/field tests and providing an access point for UTM operational data.

Graceful degradation: Designing for bounded performance under faults (maintaining safety margins and controllability rather than hard shutdown), typically via RTA, reconfiguration, or fallback modes; this is a practical expression of resilience engineering in flight systems.

GRC: Baseline ground risk class of the operation

LAANC (Low Altitude Authorization and Notification Capability): The automated system (via approved USS apps) that provides near-real-time airspace authorizations in controlled airspace at or below published grid altitudes.

OSO: Objectives (with robustness levels) one must meet to control the risk. (See JARUS SORA materials and its Annex E on link/containment robustness.)

Resilience (systems/cyber-resilience): The capacity to anticipate, withstand, recover from, and adapt to disruptions (including cyber) across the system life cycle.

Safety-I vs. Safety-II: Safety-I views safety as reducing what goes wrong; Safety-II (resilience engineering) emphasizes the ability of systems/organizations to succeed under varying conditions by anticipating, adapting, and sustaining operations.

SAIL: The resulting assurance level driving the rigor of evidence/mitigations.

SORA (Specific Operations Risk Assessment): The JARUS methodology to assess UAS operational risk and specify mitigations/assurance. It derives GRC (Ground Risk Class) and ARC (Air Risk Class), then determines SAIL (Specific Assurance & Integrity Level) and applicable OSOs (Operational Safety Objectives). Use v2.x materials for current projects.

State of Charge (SoC): The remaining usable energy in a battery (often as a percentage). NASA and battery literature define SoC in terms of usable capacity vs. energy consumed; DoD/DoE/industry use is consistent.

State of Health (SoH): The condition of a battery vs. its "as-new" state (often quantified via maximum charge vs. rated capacity), important for mission reliability estimates and reserves.

Thermal runaway: A self-heating failure mode where lithium-ion cells can overheat and vent/fire due to defects, damage, overcharge, or abuse.

Urban Air Mobility (UAM): The urban subset of AAM operations (typically up to tens of miles) using air taxis and drones in and around cities; often treated by NASA as part of the broader AAM mission.

U-space (EU): Europe's digital services framework for drone traffic management (the EU analogue to UTM), implemented via USSPs and common services for safe BVLOS scaling.

USS (UAS Service Supplier): Industry services that connect operators with UTM functions (planning, strategic deconfliction, information support) and bridge communications with the FAA interface (NASA USS specification).

UTM (UAS Traffic Management): A collaborative ecosystem (separate from, but complementary to, traditional ATC) providing planning, authorization, information exchange, and conflict management for low-altitude UAS/BVLOS operations. FAA's definition and NASA's USS/FIMS architecture describe roles and interfaces.

Vertiport: An area of land, water, or structure used or intended to be used for the landing and takeoff of VTOL aircraft, as well as other ground handling facilities, such as battery recharging and aircraft hangarage.

Vertistop: Areas that allow only landing and takeoff of eVTOLs for embarkation and disembarkation of passengers or delivery and loading of goods. It has limited facilities that exclude other ground handling activities, such as battery recharging and aircraft hangarage. It can be considered in a similar way to a "bus stop."

VLOS/BVLOS: Visual Line of Sight (VLOS) is defined in 14 CFR 107.31; Beyond Visual Line of Sight (BVLOS) refers to operations when the aircraft is out of direct visual line of the remote pilot.

VTOL aircraft (powered-lift): Person-carrying vertical takeoff and landing aircraft with dedicated lift/thrust units. EASA's SC-VTOL-01 is the reference certification basis for these vehicles in Europe.

Index

T

U

V

W

X

Z

For Product Safety Concerns and Information please contact our
EU representative GPSR@taylorandfrancis.com Taylor & Francis
Verlag GmbH, Kaufingerstraße 24, 80331 München, Germany